AF601885

Empowering Women in Developing Countries
ICT Applications and Benefits

About The Centre

The Centre for Science and Technology of the Non-Aligned and Other Developing Countries (NAM S&T Centre) is an inter-governmental organisation with a membership of 47 countries spread over Asia, Africa, Middle East and Latin America. Besides this, 12 S&T agencies and academic/research institutions of Bolivia, Botswana, Brazil, India, Nigeria and Turkey are the members of the S&T-Industry Network of the Centre. The Centre was set up in 1989 to promote South-South cooperation through mutually beneficial partnerships among scientists and technologists and scientific organisations in developing countries. It implements a variety of programmes including international workshops, meetings, roundtables, training courses and collaborative projects and brings out scientific publications, including a quarterly Newsletter. It is also implementing 6 Fellowship schemes, namely, NAM S&T Centre Research Fellowship, South Africa Mineral Processing Training Fellowship, Joint NAM S&T Centre – ICCBS Karachi Fellowship, Joint CSIR/CFTRI (Diamond Jubilee) - NAM S&T Centre Fellowship, Joint NAM S&T Centre – ZMT Bremen Fellowship and Research Training Fellowship for Developing Country Scientists (RTF-DCS) in Indian institutions. These activities provide, among others, the opportunity for scientist-to-scientist contact and interaction, training and expert assistance, familiarising the scientific community on the latest developments and techniques in the subject areas, and identification of technologies for transfer between member countries. The Centre has so far brought out 64 publications and has organised 95 international workshops and training programmes.

For further details, please visit www.namstct.org or write to the Director General, NAM S&T Centre, Core 6A, 2nd Floor, India Habitat Centre, Lodhi Road, New Delhi-110003, India (Phone: +91-11-24645134/24644974; Fax: +91-11-24644973; E-mail: namstcentre@gmail.com; namstct@bol.net.in).

Empowering Women in Developing Countries
ICT Applications and Benefits

— *Editors* —
Finarya Legoh
Suman Kapur

CENTRE FOR SCIENCE & TECHNOLOGY OF THE NON-ALIGNED AND OTHER DEVELOPING COUNTRIES (NAM S&T CENTRE)

2015
DAYA PUBLISHING HOUSE®
A Division of
ASTRAL INTERNATIONAL PVT. LTD.
New Delhi – 110 002

Cataloging in Publication Data--DK
Courtesy: D.K. Agencies (P) Ltd. <docinfo@dkagencies.com>

"Empowering Women in Developing Countries through Information and Communication Technologies" (Conference) (2013 : Solan, India)
Empowering women in developing countries : ICT applications and benefits / editors, Finarya Legoh, Suman Kapur.
pages cm
Includes bibliographical references.
ISBN 978-93-5130-668-9 (International Edition)

1. Women in development--Developing countries--Congresses. 2. Women in development--India--Congresses. 3. Information technology--Social aspects--Developing countries--Congresses. 4. Information technology--Social aspects--India--Congresses. 5. Information technology--Economic aspects--Developing countries--Congresses. 6. Information technology--Economic aspects--India--Congresses. 7. Technology and women--Developing countries--Congresses. 8. Technology and women--India--Congresses. 9. Women--Effect of technological innovations on--Developing countries--Congresses. 10. Women--Effect of technological innovations on--India--Congresses. I. Legoh, Finarya, editor. II. Kapur, Suman, editor. III. Centre for Science and Technology of the Non-Aligned and Other Developing Countries.organizer. IV. Title.

DDC 305.42091724 23

Centre for Science and Technology of the Non-Aligned and Other Developing Countries (NAM S&T Centre)
Core-6A, 2nd Floor, India Habitat Centre, Lodhi Road,
New Delhi-110 003 (India)
Phone: +91-11-24644974, 24645134, Fax: +91-11-24644973
E-mail: namstct@gmail.com
Website: www.namstct.org

Published by : **Daya Publishing House®**
A Division of
Astral International Pvt. Ltd.
– ISO 9001:2008 Certified Company –
4760-61/23, Ansari Road, Darya Ganj
New Delhi-110 002
Ph. 011-43549197, 23278134
E-mail: info@astralint.com
Website: www.astralint.com

Laser Typesetting : **Classic Computer Services**, Delhi - 110 035

Printed at : **Thomson Press India Limited**

PRINTED IN INDIA

Foreword

It is really an honor that I could contribute a few words in the publication of a book of "Empowering Women in Developing Countries through Information and Communication Technologies". In this occasion, allow me to share my expression that the publication of this book is a kind of remembrance, a wonderful idea for expressing appreciations and sincere gratitude for the excellent science and technology (S&T) development as well as relations among countries under the Centre for Science and Technology of the Non-Aligned and Other Developing Countries (NAM S&T Centre).

It is my pleasure to learn that many of us realize that science and technology (S&T) are important in every aspect of human life, as one of the factors to support the development of economic growth of a nation. They are the instrument and mechanism for continuing progress in standard expectation of living. In this sense, spreading knowledge and public awareness of S&T has become a high priority, in order to introduce policies, innovation, new technologies, etc.

The advent of information and communication technology (ICT) now-a-days provides wide range of possibilities and play as a key role on disseminating information. ICT has become convenient tool for adopting various requirements. Creative approaches using ICT applications are developed to attract segmented public in particular areas. Internet has also attraction and superiority, since the access is available all the time as well as space efficiency.

I am glad that I could take a part in maintaining the harmonized S&T development, especially in the field of ICT among developing countries, in particular for women empowerment. Indeed, many achievements have been obtained, although the condition is different from one country to another, depending on the readiness of S&T development. I believe that we can maintain a good network because we have strong spirits and motivation in utilizing ICT in S&T development. All of our great efforts would be aimed for improving each of our national developments and the quality of human lives.

My reflection also requires certain consideration to the years of outstanding network and relation. The articles in this book reflects what we have done and what we want to accomplish as developing countries, essentially in the field of empowering women in ICT. Despite of what has been greatly achieved for the past few years, the need to have evaluations and introspections to our scientific alliances are also highly demanded, since such results would be definitely important for all countries.

I believe that more fruitful collaboration should be produced, especially in the capacity building programmes (scholarship, training, joint workshop, and exchange of scientist). Such investment in the developing countries cannot be neglected, since it is for our anticipation in tackling the global challenges as well as our responsibilities to the future generation.

The Directorate General for Informatics Applications through the Directorate of Informatics Empowerment has also several programmes related to women's empowerment programmes such as "Kartini Next Generation" and other programme of empowering housewives to be ICT entrepreneurs.

Finally, I should close my remarks by confirming that the Directorate General for Informatics Applications of the Ministry of Communication and Information Technology, the Republic of Indonesia, would support the current collaboration among the Non-Aligned and Other Developing Countries.

Bambang Heru Tjahjono

Director General for Informatics Applications of
Ministry of Communication and Information Technology,
Republic of Indonesia

Preface

This book is an extended and updated version of the papers presented at the International Conference entitled "Empowering Women in Developing Countries through Information and Communication Technologies", which was organised at Waknaghat, Solan (HP), India during 1-3rd June 2013 by the Centre for Science and Technology of the Non-Aligned and Other Developing Countries (NAM S&T Centre) in association with the JUIT Centre for Women Studies (Department of HSS) at Jaypee University for Information Technology (JUIT), the Non-Government Organization of Rural Education and Development - India (READ-India) and the Women in Development (IWID). The conference had speakers from pan India and several neighbouring countries, namely Egypt, Saudi Arabia, Kenya, Nepal, Bangla Desh, etc. The objective of the conference was to exchange experiences of investigators from different member Countries of NAM and to share the progress made and the best practices followed by the neighbouring countries. The programme was designed to exchange experiences, including updates on available developing nations' best practices in all aspects of status of women's participation in ICT, to gain insight and deliberate on issues of ICT "gender divide". The Conference provided an opportunity to the specialists, experts and representatives from various developing nations to discuss their country specific problems and exchange ideas and expertise on the subject and share the details of the efforts being made by various governments in the region to effectively leverage Information and Communication technology ICT) for socioeconomic development thus encouraging members to build synergies between academia, government and industry. On the whole, the participants were expected to develop a broader appreciation of the current level of usage and management of ICT at the global level and the interactions of emerging ICT markets within the global economic debate. In attendance were 46 specialists including 22 senior officials from research, government and the private sector from 21 countries, including 17 papers presented by the Indian participants.

It was noted that developing countries have made efforts to indigenize the use of ICT albeit with different effects for their individual national programmes, given their different levels of socio-economic development and starting points. It was also observed that the conference that the low level of technological development and uptake in most of the developing countries presented a situation that could explain these countries' inability to fully exploit the benefits of the provisions enshrined in most international treaties that make certain exceptions to developing countries and the stake holders failure to successfully utilize the provision in their national ICT policies and implementation programmes. The objectives of the Conference and the publication of technical papers is to:

1. Understand critical gender equality issues related to ICT in developing countries.
2. Discuss the importance of ICT in empowering women in developing countries and to explore potential opportunities for women's empowerment.
3. Understand reasons for under-representation of women in ICT sector in the context of social and institutional barriers.
4. Emphasize on the roles and responsibilities of corporate and academic institutions to related disciplines.
5. Overview of government policies on ICT integration in women's education and employment.
6. Highlight the role of NGOs in addressing issues related to women empowerment through ICT with special focus on rural women.
7. Focus on research, strategies and tools to address the gender digital divide in national and international contexts.

This publication brings together 22 technical papers and country status reports from 13 countries These technical papers demonstrate the varying levels of application of ICT in the participating countries and potential and/or actual impact of the ICT applications and benefits on their respective mainstream economies, and overall upliftment of the social-economic developments, especially for women. A number of challenges experienced in implementing ICT applications and use by masses were sighted. Resource persons served to provide insights to the current national attempts to educate people in the use of various ICT applications and services including institutional arrangements to increase awareness and skills required for large scale use, creation and design of ICTs especially among women. The technical papers have been divided into four category of themes : (I) Critical Gender Equality Issues related to ICT in Developing Countries, (II) Importance of ICT in Empowering Women in Developing Countries – Exploring Potential Opportunities for Women's Economic, Social, Health and Political Empowerment, (III) Role of NGOs in addressing Issues related to Women Empowerment through ICT with Special Focus on Rural Women, and (IV) Sharing and Analysing Best Practices on Gender Equality and ICT.

Papers presented by the nearly two dozen resource persons highlight the needs, government and non-government bodies' efforts and some best practices and success stories across the 13 countries represented in this conference. The presentations covered the role of ICT in academic institutions in Egypt for supporting the building ICT careers. The message was that Governments, academics, IGOs, civil society, scientists (men and women) and all other stakeholders need to be armed with tools and skills to help them share and build scientific and technological knowledge and capacity for achieving sustainable development.

The Gambian governments Vision 2020 to transform Gambia into a technologically advanced and information rich society is a distinct example of efforts made by governments. Gambia's Information and Communication Infrastructure (NICI) policy and plan, the ICT4 Development (ICT4D) Action Plan, the e-government strategy and the enactment of the Information and Communication Act 2009 provided a strong foundation conducive to an ICT-led environment beneficial to the socio-economic wellbeing of the country and its people, spanning all areas such as telecommunications, internet, information technology and television/radio.

Several studies from India highlighted the role and impact of ICT in several areas that impact social and individual development. ICT has been successfully used for increasing awareness in the adolescent youth in India emphasized the importance of an individual's role both in disease prevention and health promotion and use of media to generate an understanding of the importance of ICT in women's empowerment by drawing from examples of communication on issues such as climate change. Other studies highlighted the role of ICT in the life of rural women to surmount disadvantages to learn, enjoy better health, prosper in livelihoods, access legal help and take care of government and private transactions. It was appreciated that there is an abundance of pilot projects using ICT to engage rural communities with active role of NGOs in India in spreading new ICTs to empower rural women having little formal education. Use of versatile platforms such as mobile phone applications, multi-purpose village kiosks, and interactive voice response systems and identification of challenges and recommendations to accelerate use of these technologies for empowerment of rural women was discussed at length. It was concluded that application of ICT can bring about a positive change in this regard with the help of several technological innovations namely, Internet, mobile phones, social media, radio and television etc. which help in women empowerment and providing adequate social security to women in all spheres of life through capacity building, affordable health services, employment opportunities, and improved social, economic and gender equality. The gender divide was also described in the context of the so called ICT enabled call centers in India where the working women face unique challenges both on the domestic and the work front.

Common Barriers faced by Women's Participation in ICT in S. Africa were highlighted inspite of South Africa being one of the most progressive constitutions in the world for guaranteeing equality of all its" citizens. A paper from Sudan highlighted the absence of women in higher research positions.

A few best practices were the Telecenter model of ICT Malawi: role for empowering rural women. The Affordable use of ICTs in Africa is critical, not for the mere ability to conduct social intercourse by electronic means, or to improve the delivery of governmental and business services to isolated communities. Women's access to and use of ICT is constrained by factors that go beyond issues of technological infrastructure and the government's efforts to bridge the gender gap in high performance computing.

Another example is the role of Indian government to bridge the digital divide was represented by Mission Convergence (MC) for infusing new governance architecture for actively engaging citizens with a focus on increasing women's participation. The expanded use of technology in a Public-Private-Partnership (PPP) model enabled the system to include poor and vulnerable women living in slums and other underserved areas, in the ambit of Government's welfare programmes and other public services. Gender Resource Centre (GRC) - *Suvidha Kendras* helped overcome the socio-cultural barriers hitherto precluding women's equal access to opportunity in the areas such as health, education, skill and social protection provisions.

The participatory communication initiatives in India by READ centers are yet another example of a best practice. They use varied tools, the knowledge centers and the community radio for reaching women with locally relevant content. The content in local and simple language, enables everyone in the community, including women, who can then understand and make the best use of technology emphasizing the need for initiation of a knowledge revolution through effective and meaningful use of information and communication technologies.

Role of Indonesian ICT volunteers for the development of Batur Geopark web site at Bali is a striking example of ICT mediated promotion of various disciplines achieving synergy through informal science education, journalism, festivals, promotion, and these make the implementation of programmes is more knowledgeable.

The 1nita and the Internet Entrepreneurship Programme of the Malaysian government provide a sustainable platform for women entrepreneurs and women-owned SMEs to build business capacity for the future. This project my Domain Registry, an agency under MOSTI aims to empower women in achieving long-term business success by taking advantage of technological developments and the Internet.

An active role has been played by the government and non-government organizations in Myanmar to enable women to participate fully in the national development and ICT education programmes regardless of nationality, race or religion.

A framework has been developed for use of ICT in rural Kenya for decreasing poverty especially among Kenyan women. This framework guides gender responsive mobilization and utilization of resources for effective engagement of women in all public/private sectors and bring women out of the 'darkness', shed off illiteracy bringing about a paradigm shift from the traditional 'analogue' state to the current 'digital migration' by both the elite and the transformed illiterate.

Sudan has tried to use ICT in reducing the gap in opportunities for women in a country such as Sudan, which is bereft with conflicts forgoing the need of the people to basic services, the restricted humanitarian aid provided by donors due to the US sanctions, corruption of the system and the limited support extended by the private sector in the form of philanthropy. ICT based spread of information is increasing awareness about environmental awareness for restoring plant cover of the area, used alternative energy source safely to combat desertification in Sudan.

ICT has been used for documentation and in association with Zambia National Broadcasting Corporation is preserving Zambia's Intangible cultural heritage by collecting and stories, for use by the Zambian communities. This attempt establishes the importance of stories as well as the use of local languages; show the value of storytelling in modern cities like Lusaka and encourage further research on the importance of folktales of other ethnic groups in Zambia.

During the discussion of the conference, it was identified that a "gender divide" was persistent and led to lower numbers of women accessing and using ICT compared with men. There was tenacious gender discrimination in labour markets, education and training opportunities, as well as allocation of financial resources for entrepreneurship and business development. The limitation of infrastructure and technology development in most of the cases presented situation that explain the countries' inability to fully exploit the benefits of the necessities of the specific programmes. In conclusion the workshop noted the challenges cited by most of the country presenters included; absence of supportive institutions, lack of basic infrastructure (insecurities in power, water, general communication facilities and sometimes food and feed in times of extreme weather events); inadequate or fragmented domestic ICT related legislation; absence of locally available ICT capacity building and training programmes. It was also noted that developing countries have made their efforts to emphasize, explore, and develop potential opportunities for women's participation in ICT in supporting development in economic, social, health and political sectors. Most of the preliminary programmes were considered carefully to match the particular situation and condition of the targeted women and associated areas, as well as the culture and behaviour of the locals and there is optimism to apply ICT as a powerful driver in improving women empowerment. However, developing countries have to pace up their efforts in promoting ICT use as this projects rapid change for the economic development paradigm premised on future knowledge, and pose a challenge that requires urgent action in all forms – skills development, curriculum, new industries including moral and ethical issues all providing an indication of increased competitiveness on the rapidly expanding global technology applications. It is anticipated that ICT will open channels of equality for ways of working, interacting and learning, to enhance the living conditions and opportunities for the developing nations around the globe.

This book is a valuable resource for gaining information and knowledge about how the developing countries are addressing the issues of "gender divide" with various policies, approaches and significant programmes, which are alike and comparable from one another.

Finarya Legoh

Principal Engineer
The Agency for Assessment and
Application of Technology (BPPT)
Jakarta
Indonesia

Suman Kapur

Senior Professor and Dean
International Programmes & Collaborations Division
Birla Institute of Technology and Science
(BITS Pilani) Hyderabad Campus
Hyderabad – 500 078, India

Introduction

Information and Communications Technology (ICT) has dramatically changed the world. It is a powerful driver that has enhanced the living conditions and opportunities for the nations around the globe. The ICT sector has become a key economic factor for the developed as well as the developing countries alike and has opened the channels for new ways of working, interacting and learning for both men and women. It has however been observed that the involvement of women in ICT sector is highly inadequate, particularly in the developing nations and there is a 'gender divide' resulting in lower numbers of women accessing and using ICT compared with men. Unless this problem is specifically addressed, there is a risk that ICT may exacerbate existing inequalities between women and men and create new forms of inequality. The existing persistent gender discrimination in labour markets, education and training opportunities and allocation of financial resources for entrepreneurship and business development negatively impact women's potential to fully utilise ICT for economic, social and political empowerment. If the gender dimensions of ICT in terms of access and use, capacity building opportunities, employment and potential for empowerment are explicitly identified and addressed, these can be a powerful catalyst for economic, political and social empowerment of women, and the promotion of gender equality.

In order to deliberate on these issues, the Centre for Science and Technology of the Non-aligned and Other Developing Countries (NAM S&T Centre) and the JUIT Centre for Women Studies, Jaypee University of Information Technology (JUIT), Waknaghat-Solan, India organised an International Conference on 'Empowering Women in Developing Countries through Information and Communication Technologies' at JUIT-Waknaghat, District Solan (HP), India, during 1-3 June 2013. Initiatives: Women in Development (IWID) and Rural Education and Development (READ) - India were the collaborators of this scientific event.

46 specialists from 20 countries, namely, Angola, Egypt, The Gambia, Indonesia, Iraq, Kenya, Malawi, Malaysia, Mauritius, Myanmar, Nepal, Nigeria, South Africa,

Sri Lanka, Sudan, Togo, Uganda, Venezuela, Vietnam and the host country India participated in this International Conference. Altogether 39 scientific papers and country status reports were presented during the Conference - 22 by the foreign participants and 17 from the Indian participants.

The present book is an edited compilation of 19 papers read during the Conference and found suitable for this publication. At the end of the book is placed a copy of the Waknaghat Recommendations – 2013 on 'Empowering Women in Developing Countries through Information and Communication Technologies' that were unanimously adopted by the Conference participants.

I gratefully acknowledge the dynamic involvement and untiring effort of Dr. Finarya Legoh, Principal Engineer of the Agency for Assessment and International Programmes and Collaborations Division Application of Technology (BPPT) of Indonesia and Prof. Suman Kapur, Senior Professor and Dean at the Birla Institute of Technology and Science (BITS Pilani), Hyderabad Campus of India for technical editing of this publication. Last but not the least, valuable services provided by the entire team of the NAM S&T Centre, in particular, Mr. M. Bandyopadhyay for overall supervision, and Ms. Shania Tahir and Mr. Pankaj Buttan in compiling the papers, coordinating between the editors and the authors and giving shape to this volume are greatly appreciated.

I hope that the information provided and suggestions and recommendations made in the papers included in this book will help the policy makers for accelerating the process of empowering women through ICT in the developing countries.

Prof. Dr. Arun P. Kulshreshtha

Director General,

NAM S&T Centre

Contents

Foreword *v*
Bambang Heru Tjahjono (Indonesia)

Preface *vii*
Finarya Legoh (Indonesia) *and Suman Kapur* (India)

Introduction *xiii*
Prof. Dr. Arun P. Kulshreshtha (India)

Part I
Importance of ICT in Women Empowerment

1. **Empowering Young Women in Developing Countries through ICT: A Public Health Approach** 3
Suman Kapur (India)

2. **Empowering Science and Technology Communication through Information and Communication Technology: Programmes in Indonesia** 13
Finarya Legoh (Indonesia)

3. **The Roles and Responsibilities of Academic Institutions to Encourage Women Participation in ICT Related Disciplines** 27
Shireen K. Assem (Egypt)

4. **Empowering Rural Women for Socio-economic Development through Information, Communication and Technology: Challenges and Opportunities for Malawi** 37
Catherine Chaweza (Malawi)

5. **Women Empowerment through Information and Communication Technology in Nepal** 57
Parbati Pudasaini (Nepal)

6. **Social Security of Women in Developing Countries through ICT** 67
Isha Sehgal, Pinky Singh and Parul Sehgal (India)

Part II
Important Role of NGOs in Women Empowerment through ICT with Special Focus on Rural Women

7. **ICT Framework Used as Women's Tool for Poverty Alleviation in Rural Areas of Kenya** 99
Catherine Kathure Kaimenyi (Kenya)

8. **ICT and Women's Empowerment: A Paradigm Shift Needed in Outlook** 113
Archita Bhatta (India)

9. **Harnessing the Power of New ICTs for Rural Women in India: NGO Roles** 121
Jane Schukoske (India)

10. **Women's Empowerment Gap and Sustainable Development in Sudan** 139
Hanan Mohamed A. Karim Abbas (Sudan)

11. **Myanmar Women and Girls in ICT** 157
Khin Mar Lar Tun (Myanmar)

12. **Empowering Women Skills in Combating Desertification in Sudan** 165
Maha Ali Abdel Latif and Eiman Elrashid Diab (Sudan)

Part III
Best Practices and Case Studies on Gender Equality and ICT

13. **Country Status Report on ICT Development in Gambia** 179
Bintou Dibba (The Gambia)

14. **Information Technology for Women Empowerment: The Case of *Suvidha Kendras* under Mission Convergence** 185
Rashmi Singh (India)

15. **Empowering Women in Developing Countries through Information Communication Technologies** 205
Geeta Malhotra and Yashpal Malik (India)

16. **Women's Empowerment through ICT: Case Study of 2 Projects under Ministry of Science, Technology and Innovation (MOSTI), Malaysia** 219

Puspadevi Kuppusamy (Malaysia)

17. **Women in Research and Development Institutions: Case Study of National Center of Research** 233

Eiman Elrasheed Eltayeb Diab, Migdam Elsheikh Abdel Gani, Maha Ali Abdel Latif and Gamma AbdELgader Osman (Sudan)

18. **Globalizing India: The Sunshine Sector and its Shadows– The Call Center Industry: A Case Study** 241

Rekha Pande (India)

19. **High Performance Computing in Materials Science Research: Are There Women?** 261

Rapela Regina Maphanga (South Africa)

Annexure–
Waknaghat Recommendations - 2013 on Empowering Women in Developing Countries through Information and Communication Technologies 273

Part I

Importance of ICT in Women Empowerment

Chapter 1

Empowering Young Women in Developing Countries through ICT: A Public Health Approach

Suman Kapur

Dean, Research and Consultancy,
Faculty-Incharge Education Development Division and Professor,
Biological Sciences Department,
Birla Institute of Technology and Science, BITS-Pilani,
Hyderabad Campus, Hyderabad, India

ABSTRACT

Decision making is a vital element in health seeking behaviour of women. The freedom to make a right choice and access to health care (access in terms of need) can be enhanced through Information and Communication Technologies (ICT). Technologies can break down barriers to health information and can build women as primary users and disseminators of the same. ICT can impede violence which is critical in affecting women's health in developing countries. For a rural woman from developing country, the 'digital divide-inequality in ICT' amounts from two sources. Firstly the limited access to resources and facilities due to poverty, illiteracy, lack of computer literacy and language barriers, and secondly the 'gender divide' which impedes women from utilizing the facilities due to gender discrimination (1). This inequality needs to be tackled effectively to lead the rural women into healthy and empowered life.

One of the strategic objectives of Beijing Declaration and Platform for Action adopted at the Fourth World Conference on Women in 1995 was to increase the participation and access of women to expression and decision-making, in and through the media along with the use of new technologies for communication (2). The information revolution among women was further emphasized afterwards through UN commission on status of women, Economic and Social Council and World Summit on the Information Society (3-6). ICT application for e-health was put forward by first phase plan of action of World Summit on the Information Society along with special needs of women in relation to capacity-building, enabling environment, cultural diversity and identity, media, follow-up and evaluation. The global and regional response to this call has initiated many success stories (7, 8). Among the 250 million households in India, 63.2 per cent have telephones which makes the absolute number of mobile subscribers around 564 million, which is enough to serve nearly half of the country's 1.2 billion population (9). Thus there is a wide scope of making use of media and telecommunication system to empower women in rural India.

Gender discrimination replicates itself from generation to generation, violates the rights of the girl child and chokes her overall development. The aim is to reduce gender discrimination and improve health of adolescent girls in Andhra Pradesh. It proposes to look into the health risks which stem from a combination of socio-cultural, economic and heath care related factors. Socio-cultural customs such as early age of marriage and consequently child-bearing at a young age places young women at heightened risk for reproductive morbidity. Poor health information available to adolescents especially living in rural area is also a significant factor impacting their health and well-being. Adolescence is the right time to intervene as the youth establish their patterns of behaviors and make lifestyle choices that affect both their current and future health. Overall finding points towards developing an inclusive, integrated ICT-based, youth development programme to meet the needs of adolescents, especially girls with their families and their communities will be presented.

Introduction

Role of Information and Communication Technology (ICT) is multifaceted in empowering women in developing countries. ICT can help in economic empowerment by improving women's access to income-earning, opportunities and productive assets (1). It can influence the well-being by improving access of rural women to basic services and infrastructure. Decision making is a vital element in health seeking behaviour of women. The freedom to make a right choice and to access the health care (access in terms of need) can be enhanced through ICT (2). Increasing women's say in community affairs and strengthening women producers' organizations can also be considered as a good sign of empowerment. ICT can impede violence which is critical in affecting women's health in developing countries. Technologies can break down barriers to health information and can build women as primary users and disseminators of the same.

For a rural woman from developing country, the 'digital divide-inequality in ICT' amounts mainly from two sources. Firstly the limited access to resources and facilities due to poverty, illiteracy, lack of computer literacy and language barriers as a citizen of developing nation, and secondly the 'gender divide' which impedes women from utilizing the facilities because of gender discrimination (3). This inequality needs to be tackled effectively to lead the rural women into healthy and empowered life. The objective of this paper is to look into the international strategic and advocacy approaches for empowerment of women through ICT and how India responded to the call with examples of some case studies.

International Strategies in Empowering Women through ICT

World Conference to appraise the achievements of the UN Decade for Women was held in Nairobi, Kenya during July 1985. The Conference put forward the Nairobi

Forward-looking Strategies for the Advancement of Women to the year 2000 (4). In preparation for the Nairobi Conference, the Director General of the World Health Organization (WHO) issued a report analyzing the situation regarding women health and development; and drawing attention to the special health needs of women as well as to the key roles that women play in promoting health and development. Later down the time lane international agencies recognized the role of ICT and strategies were put forward for the Fourth World Conference on Women in Beijing.

One of the strategic objectives of Beijing Declaration and Platform for Action adopted at the Fourth World Conference on Women in 1995 was to increase the participation and access of women to expression and decision-making, in and through the media and new technologies of communication (5). The information revolution among women was emphasized afterwards through the UN commission on status of women, the Economic and Social Council, the World Summit on the Information Society (6-8). The ICT application for e-health was put forward by first phase plan of action of the World Summit on the Information Society along with special needs of women in relation to capacity-building, enabling environment, cultural diversity and identity, media, follow-up and evaluation. The advocacy approaches on women empowerment through ICT included the Division of Advancement of Women (Women 2000), the United Nations Inter-Agency Network on Women and Gender Equality (IANWGE), and the World Summit on the Information Society (WSIS) (9).

Case Studies from India on ICT and Women Empowerment

The global and regional response to this call has initiated many success stories (10, 11). In India there were action plans put forward through Government schemes, NGOs and research institutions. One of the early ventures included the Village Knowledge Centers (VKC) by M.S. Swaminathan Research Foundation, where telecommunications infrastructure and community based innovations are carried out through women volunteers in Tamil Nadu and Union Territory of Puducherri since 1998. The VKC was the instrument in providing educational services, employment and income generation, skill development, weather and specific interventions through fisherwomen volunteers. It also provided services through linkage and partnership including health services, animal husbandry, data base on BPL, etc. The reported impact of the same included economic and knowledge empowerment of women, enhanced decision making ability and self-esteem among rural women, weakened customary social exclusion (towards Dalits) and improved social awareness and participation towards government schemes. Women proved their competence in managing the VKC and gained respect in community and family. It gave enhanced access to health services for women and children through partnership programmes.

There were few state level governmental programmes like Dr. SMS-Mobile health information system in Kerala since 2008 (12). The purpose was to provide comprehensive information on health-related resources via the short message service (SMS). It was opened exclusively for m-services of the state as a pilot study and now it is backed by a database of health infrastructure in 10 out of 14 districts in the state. The facility provides timely information on medical facilities available in that locality including contact, phone number, specialties and super specialties (Cardiology, Pediatrics, Ophthalmology), doctors, ambulance services, etc. The project also incorporates a health portal and web based information mechanism enabled through GIS.

In the national level, the Health Information System Programme (HISP India) has made an impact through partnering with state governments using multi-disciplinary approach and seeking to the knowledge domains of public health and informatics (13). The timely availability of health information helps the health workers for surveillance, monitoring and intervention on conditions like epidemics, maternal and child health problems.

A point of care device from the Public Health Foundation of India named Swasthya Slate (Health tablet) tried to incorporate multiple diagnostic tests for improving decision making particularly among rural women (14). The device stores the patient data and their location (GPS) and put the same to cloud for reference and care. This allows offline/online operations and doctor on call services. The outcome can be improved access in terms of need, decision making power with timely recommendations which would be greatly helpful for pregnant mothers and children. In this application, clinical algorithms based on customized protocols developed by PHFI is used for screening which allows users to deliver fast and accurate care at home and improve health awareness.

Self-help groups (SHGs) play an important role in economic empowerment of rural women in India (15). They are aimed at well-being and economic independence of women through self-employment and entrepreneurial development. In India more than 25 lakh of SHGs are there in the country, where 90 per cent of them are women SHGs. Reports from the states like Andhra Pradesh showed that 90 per cent of members make regular savings, where the women got access to credit and increase in income has been spent on better nutrition of children and on the health care for the family. The indicators showed that the members adopt small family norms and improved health seeking and vaccination status. Role of ICT in this progress included marketing of SHG products, quality control of products, information on animal husbandry, education, local farming, weather, maintaining accounts, payments, bank transactions, run PCOs, computer centers and television channels.

A gender responsive budget scheme from the government of India named 'Sanchar Shakti' aimed at facilitating SHGs' access to ICT enabled services and their contribution towards ICT enabling services (16). It was initiated by the Universal Service Obligation Fund (USOF) of the Department of Telecommunications (DoT) and the Ministry of Communications and IT. Project categories included the provision of subsidized mobile Value Added Services (mVAS) subscriptions to SHGs, the setting up of SHG run mobile and modem repair centers in rural areas, the setting up of SHG run solar-based mobile phone, fixed wireless and terminal centers for phone charging in rural areas. Seven pilot projects of Sanchar Shakti included the villages situated in the States of Andhra Pradesh, Kerala, Maharashtra, Tamil Nadu, Uttar Pradesh, Uttarakhand and Union Territory of Puduchchery. Through this scheme information related to wellbeing (health, social issues and government schemes), entrepreneurial activities (markets and financial products and skill enhancement) and lifestyles are given to women SGH members through SMS and IVRS. Reports from states like Rajasthan showed improved income of rural women engaged in the horticulture industry through awareness of better agricultural practices and animal husbandry. The scheme also played a role in overall empowerment of women through information on health care, education, women upliftment and government schemes.

The Self-Employed Women's Association (SEWA) made use of ICT for empowerment of rural women in Gujarat and Bihar through innovations like ICT School for Empowerment of Women and setting up the Community Learning and Business Resource Centers CLC (Sanskar Kendras). The ICT School contributed for training of trainers in different micro enterprise development, to house a community radio to support local farmers, to provide training in employable skills and train and help women for rural entrepreneurship. CLC is an instrument in women empowerment through livelihood enhancement, business development services, supportive services and capacity building. The project covers more than 240,000 members in Gujarat and supports more than 5 organizations of Bihar (17).

ICT to Prevent Violence against Women

There are few recent advances in ICT to prevent violence against women. The Circle of 6 app for iPhone and Android is used worldwide and launched in India in April 2013 in Hindi (18). It gives quick and easy access to reach the 6 selected friends in time of need. The app is connected with GPS (location) and sends SMS for help, call or chat. An innovation named Society Harnessing Equipment (SHE) is introduced by engineering students in Chennai. It is an anti-rape innerwear laced with a global positioning system (GPS) and a global system for mobile communications (GSM), which can transmit messages to the owner's family members (19). It can also emit shock waves to deter potential assailants. FightBack is also a women's safety application, sends SOS alerts from phone. It uses GPS, SMS, location maps, GPRS, email and Facebook account, to inform connected people in case of danger. All the reports are updated on a map in real-time to show "harassment hotspots" and dangerous areas (20).

What is Missing?

There are few areas that still need to be addressed through ICT for overall empowerment of women. One of the key areas includes adolescent health, where health issues of young women need to be intervened through a confidential and comfortable approach. ICT can tackle the limitations of conventional face to face counseling and help-line approach. The scope of Innovations with ICT include Cyber-counseling (Anonymous form of counseling) where anonymity can make communication easier through the Internet for young people (21). This method can discuss more severe and complex concerns of child abuse, suicide and mental health problems. Another option is providing young adults with online cloud for building awareness related to development and sexual health where set of frequently asked questions and knowledge tips is accessed. But these innovations need to be reviewed and pilot tested for challenges including therapeutic issues like lack of nonverbal messages (need research on how to convey emotional clues in text based messages), ethical issues to ensure proper care, and legal issues to manage if something goes wrong.

Another future scope of ICT is life course in health counseling as needed for age specific health counselling for women is evident from recent research. ICT can support women's health in milestones like puberty, pregnancy, child birth and child-rearing, family planning, menopause and elderly care. It can also provide rural women with decision matrix (framework) in health care. Cyber polling (online voting) is a platform for advocacy for women empowerment by giving feed-back to policy makers and building a feeling of homogeneity, to bring to the attention of community. A fine example is the open letter prepared by the Amnesty International to vote for a new law on violence against women in India, initiated after the rape and death of a 23-year-old student in Delhi in December 2012 (22).

India in the 'Last Mile Connect'...!

India is recognized as a giant in the global IT map for export-oriented software and ICT based services sector. But the ICT has failed in providing broader development and benefits to all Indian citizens. Among the 250 million households in India, 63.2 per cent have telephones which makes the absolute number of mobile subscribers around 564 millions, which is enough to serve nearly half of the country's population of 1.2 billions (23). This shows a wide scope of making use of the media and telecommunication system to empower women in rural India (Table 1.1).

Table 1.1: A Matrix for the Future Scope of ICT on Women Empowerment (summarized from 24)

Strategies to Tackle the Digital Divide	*Support from Policy Makers*	*Research and Evaluation*	*Entrepreneurs and Gate Keepers*
Scale up initiatives to have effects throughout India	Universal service programmes to bring connectivity	For understanding of both positive and negative unintended consequences of mHealth and eHealth intervention for women	Need more entrepreneurs to bring ICT accessible to rural women
Strategy for attitudinal and institutional change	Relevant content to make connectivity meaningful	Research on gender based violence and ICT (influence of media)	Involvement of men and other key gatekeepers who influence women's decisions
Integration in wider socio-technical interventions	Capacity building of state, and private players is essential to use content and facilitate its use by rural women.		

Figure 1.1 shows the '10 point action plan' prepared, based on the recommendations and background information. The most important key initiative is ICT accessibility to rural women with more women entrepreneurs. This is expected to

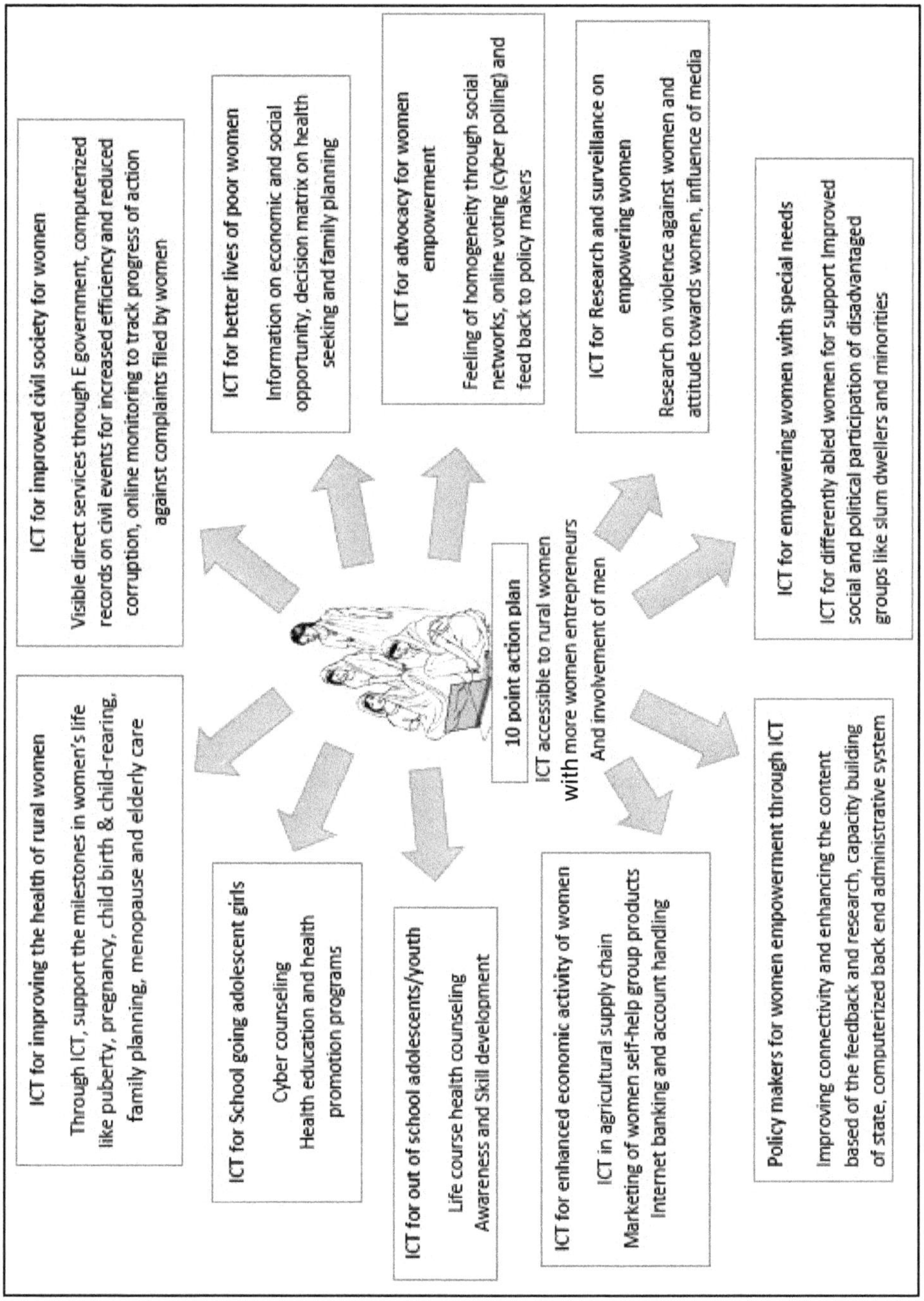

ICT for improving the health of rural women

Through ICT, support the milestones in women's life like puberty, pregnancy, child birth & child-rearing, family planning, menopause and elderly care

ICT for improved civil society for women

Visible direct services through E government, computerized records on civil events for increased efficiency and reduced corruption, online monitoring to track progress of action against complaints filed by women

ICT for School going adolescent girls

Cyber counseling
Health education and health promotion programs

ICT for better lives of poor women

Information on economic and social opportunity, decision matrix on health seeking and family planning

ICT for out of school adolescents/youth

Life course health counseling
Awareness and Skill development

ICT for advocacy for women empowerment

Feeling of homogeneity through social networks, online voting (cyber polling) and feed back to policy makers

10 point action plan

ICT accessible to rural women with more women entrepreneurs And involvement of men

ICT for enhanced economic activity of women

ICT in agricultural supply chain
Marketing of women self-help group products
Internet banking and account handling

ICT for Research and surveillance on empowering women

Research on violence against women and attitude towards women, influence of media

Policy makers for women empowerment through ICT

Improving connectivity and enhancing the content based of the feedback and research, capacity building of state, computerized back end administrative system

ICT for empowering women with special needs

ICT for differently abled women for support Improved social and political participation of disadvantaged groups like slum dwellers and minorities

Figure 1.1: 'Ten Point Action Plan' for Empowering Rural Women through ICT.

provide the framework for the comprehensive plan with focus on health, education, economic and social activities.

Conclusions

The global community has sensed the importance of ICT in empowerment of women over the years through the remarkable initiatives from governmental and non-governmental organizations. The initiatives included are both a part of development policies and projects. The innovations and strategies need to ensure that the ICT revolution happening round the globe should not bypass the rural women. Platforms like the International Conference on Empowering Women in Developing Countries through Information and Communication Technologies should act as a catalyst in bringing researchers, academicians, innovators and policy makers to address the gender digital divide in national and international contexts.

References

1. ILO. Skills for improved productivity, employment growth and development Geneva: International Labour Conference, 97th session; 2008. Report No.: V. http: //www.ilo.org/wcmsp5/groups/public/@ed_norm/@relconf/documents/meetingdocument/wcms_092054.pdf
2. VAPS. Enhancing Women Empowerment through Information and Communication Technology. Voluntary Association for people service; 2005. http: //wcd.nic.in/research/ict-reporttn.pdf
3. UN. Gender equality and empowerment of women through ICT, Women 2000 and beyond, September 2005 http: //www.un.org/womenwatch/daw/public/w2000-09.05-ict-e.pdf
4. UN. Report of the World Conference to review and appraise the achievements of the United Nations Decade for Women: Equality, Development and Peace. New York: United Nations; 1986. http: //www.un.org/esa/gopher-data/conf/fwcw/nfls/nfls.en
5. UN. The Fourth World Conference on Women. Beijing; 1995 http: //www.un.org/esa/gopher-data/conf/fwcw/off/a—20.en
6. UN, Women 2000 - Women and the Information Revolution, New York, 1996. http: //www.un.org/womenwatch/daw/public/Women2000%20-%20Women%20and%20the%20Information%20Revolution.pdf
7. UN, Report of the Ad Hoc Committee of the whole of the twenty-third special session of the General Assembly, Suppliment No. 3 (A/S- 23/10/Rev.1), New York, 2000 http: //www.un.org/womenwatch/daw/followup/as2310rev1.pdf
8. UN. Development and international cooperation in the twenty-first century: the role of information technology in the context of a knowledge-based global economy. UN Economic and Social Council; 2000 http: //www.un.org/documents/ecosoc/docs/2000/e2000-l9.pdf
9. UN Women. Commission on the Status of Women-Follow-up to Beijing, 47th Session 2003 http: //www.un.org/womenwatch/daw/csw/47sess.htm.

10. Health Net https: //www.healthnet.com/portal/home.do
11. FHI360-SATELLIFE Health Information and Technology http: //www.healthnet.org/
12. GoI. Dr. SMS: Introducing healthcare through mobile technology in Kerala Governance Knowledge Centre. http: //www.indiagovernance.gov.in/cms/preview_bestpractices.php?id=1523.
13. HISP India http: //hispindia.org/.
14. PHFI. Swasthya Slate: http: //www.swasthyaslate.org/.
15. PSR M. Economic Empowerment of Rural Women by Self Help Group through Micro Credit [Internet]. Rochester, NY: Social Science Research Network; 2012 http: //papers.ssrn.com/abstract=2193045.
16. GoI. DOT-USOF'S Gender Budget Scheme "Sanchar Shakti" Press Information Bureau. 2011 http: //pib.nic.in/newsite/erelease.aspx?relid=70578.
17. Sulaiman V R. ICTs and Empwerment of Indian Rural Women. CRISP, Hyderabad; http: //www.crispindia.org.
18. Circle of 6 App Circle of 6/» Circle of 6 App http: //www.circleof6app.com/.
19. New Anti Rape Gadget, SHE (Society Harnessing Equipment) | Strange Or What http: //www.strangeorwhat.com/new-anti-rape-gadget-she-society-harnessing-equipment/.
20. Fightback, India's first mobile app for women's safety http: //www.fightbackmobile.com/welcome.
21. David J. Powell. Cyber Counseling, The Growth of Online Counseling, 2012 http: //www.recoverytoday.net/articles/445-cyber-counseling.
22. Amnesty International: Vote for a new law on violence against women in India 2013 http: //www.amnesty.org/pt-br/library/info/ASA20/011/2013/en.
23. Census of India: Houselisting and Housing Census Data Highlights - 2011 http: //www.censusindia.gov.in/2011census/hlo/hlo_highlights.html.
24. Geoff Walsham. ICTs for the broader development of India: An Analysis of the literature. EJISDC. 2010;(41): 1–20.

Chapter 2

Empowering Science and Technology Communication through Information and Communication Technology: Programmes in Indonesia

Finarya Legoh

Principal Engineer,
The Agency for Assessment and Application of Technology (BPPT),
Jakarta, Indonesia
E-mail: finaryalegoh@gmail.com; finarya.legoh@bppt.go.id

ABSTRACT

As a developing country, Indonesia has set goals for its future economic, social and environmental sustainability, which depend upon the challenging Science and Technology (S&T) capability and competitiveness. The first pleasurable S&T communication was established in 1991, when Indonesia Science and Technology Centre was initiated by the Ministry of Research and Technology, supported by the Agency for Assessment and Application of Technology (BPPT).(*) The main objectives are to attract people especially youngsters to love S&T, and to encourage public engagement in S&T. Since then, related S&T communication programmes have been nurtured widely.

Although the importance of S&T is realized, S&T communication is nevertheless not a popular persistence and still a novel knowledge in Indonesia. Seed of communication has been transformed in sporadic ways as comic posters, hands-on exhibits, games, etc. The common communication is due to campaign such as science Olympiad, or demands in defeating national problems, such as H5N1 and global warming. There are still perceptions, however, that S&T communication is merely a tool in assisting education for youngsters involving S&T. A gap remains in synergizing R&D institutions, science centres, media, formal education, and S&T for everyday life.

The Government of Indonesia has put efforts and initiatives to create programmes in order to bridge the gap in S&T popularization and to educate public in S&T commonality. One of the programmes is applying information and communication technology (ICT) as a tool for S&T promotion, since the ICT convergencehas fast developed recently by means of multimedia access, mobile devices and other advance technologies. ICT has become the most useful and popular form of media available to public.

S&T knowledge should be spread out in the regions, especially where wide area of Indonesia is mostly rural. In this sense, S&T communication is noteworthy in supporting urgent issues need to be addressed. The programmes have been developed as case of practices, e.s: Indonesia ICT volunteers for the regions and the development of Batur Geopark web site at Bali. Various disciplines are synergized through informal science education, journalism, festivals, promotion, and these make the implementation of programmes more knowledgeable.

The other programme is especially created to empower women through ICT. The fast growth of Social Network Services (SNS) has attracted women to be more familiar with computer and smartphone. Inside Facebook report 2009 stated that more than 56 per cent women in the USA is Facebook audience. Unfortunately, this attraction is not accompanied by their knowledge in ICT. In some cases, this technology illiteracy brings some issues, such as data security hacked, fraud transaction, etc. The DigiMoM offers a blended format of café scientifique and workshop in promoting the ICT awareness.

ICT media could make S&T public communication more effective as a hub, and stands between scientists and public to provide suitable information through several channels. In the case of NAM, there are contemporary S&T based issues which the governments and people of NAM countries have to deal with, especially for empowering women through ICT. People should consider how they can work together to solve these problems by using scientific approaches. It will develop and enlarge the capacity to nurture the NAM S&T network.

Keywords: *Public awareness of science and technology, Science and technology communication, ICT media, Empowerment of women through ICT.*

Introduction

Science and Technology (S&T) become powerful tools for sustainable economic and social capability and competitiveness, especially for Indonesia as a developing country. The importance and challenging areas of research and development of S&T have been set in 2002 through the Law Number 18/2002 of the National System of Research, Development and Application of S&T. The substance of the Law includes:

1. Accelerate the development and application of S&T resources effectively.
2. Trigger networking among research and development institutions.
3. Regulate the inter-connection between the doers and the users of S&T.
4. Define roles and functions of government (local and central) and public to actively contribute in the participation of S&T research, development (R&D) and application.

5. Regulate tasks and responsibilities of government, R&D institutions, public, and also possibilities of sharing resources among them and others.

In this sense, spreading knowledge and public awareness of S&T has become a high priority in order to introduce policies, innovation, new technologies, etc. S&T communication has its place in the area of R&D and plays an important role as a hub between S&T and public, so that new developments of R&D can be accepted effortlessly by public.

S&T communication is actually a mediator and stands between scientist and public to provide suitable information through severals channels. It is, nevertheless, not yet popular, pervasive and is still a new approach, although the seed of activities are helping to change things. Programmes and activities were adopted by science centres, government and R&D institutions, as well as private business and foundations concerned about S&T awareness. The communication is commonly associated with campaigns such as the Science Olympiad competition, attacking H5N1, Tsunami, global warming, etc. Various supporting materials are created and distributed as comic posters, hands-on exhibits, games, etc.In this sense, role of women is important to take part in communicating the campaigns to public.

The Government of Indonesia represented gender empowerment policies and programmes mainly through the Ministry of Women Empowerment and Child Protection (PP and PA). All government institutions are recommended to have gender empowerment programmes individually or through collaborations. Besides, women empowerment is also established in various sectors, such as: the National Council of Women's Organization Indonesia (KOWANI), in which 82 women organizations have joined Kowani as the umbrella of women organizations, the Association of Business Women of Indonesia (IWAPI), Board of National Family Planning, etc.

Although various programmes of gender empowerment have been established and implemented, the institutions network need to be strengthened and synergized. Activities developed are focused to the basic problems as to attack and resolve women and children social problem in relation of gender equality and equity, *e.g.* women and child trafficking, health, illiterate, social welfare, education, etc. In national education – care is given to the education of female students especially those who came from low level families, unfinished education, and/or they are forced to work for the family. Women in S&T especially in information technology and communication (ICT), however, are not really promoted, although the Ministry of Research and Technology (MRT) and the Ministry of Communication and Information Technology (MCIT) have conducted related programmes to increase research for women scientists and engineers.

The advent of ICT provides wide range of possibilities and plays as a key role on disseminating information. ICT has been progressively developed lately and has become convenient tool for adopting various requirements. Creative approaches using ICT applications are developed to attract segments of public in particular areas. The popularization of social media services are other recent addition to the means of communcation with infinite resources from the internet. It is, therefore, that the

Indonesian Government supports applying ICT in developing programmes for communicating policies, knowledge, and new developments.

In this sense, ICT is generally dominated by men, whereas women users are more reluctant to learn the advantage of the ICT application. This ICT illiteracy could hinder women from various work advantages, as well as bring some problems such as data security mistakes, hacked and fraud transactions, etc. Women in all sectors within rural and urban areas should be encouraged, made confident, and put on a similar level as opportunity for men in applying ICT.

There are three programmes conducted by those Ministries to emphasize women empowerment in S&T that are explained later in this paper, as these programmes are examples that have been implemented through inter-sector collaborations.

Women Empowerment Institutions and Activities in Indonesia

Gender empowerment policies and programmes in Indonesia are conducted mainly by the Ministry of Women Empowerment and Child Protection (PP and PA). All government institutions are recommended to also have gender empowerment programmes individually or through collaborations.

The Vision of PP and PA is creating gender equality and child protection, whereas its mission and objectives are:

- ✰ Mission : to support/accelerate the forming of gender responsive policies and care for children to increase quality of life and women protection, as well as to fulfill the rights for growth and child protection from violence.
- ✰ Objectives:
  - Support and facilitate the forming and implementation of gender responsive policies and child care in all priority of development fields;
  - Support and facilitate the fulfillment of women and children rights and protection from violence;
  - Increase the capacity building and network and the role of society in supporting gender equality and equity, women empowerment and child protection;
  - Create accountable and integrated management for the work.

The Government of Indonesia sets the Legal Foundation and national strategy that foresee the national development from gender perspectives through Gender Mainstreaming Programme (PUG) in the central and other regions. A Decree for General Guidance of Gender Mainstreaming Implementation then is established:

- ✰ To integrate gender mainstreaming within local Government policies in planning, implementation, budgeting, monitoring and evaluation, integrated with local policies and implementation of programmes in the regions.
- ✰ To collaborate gender perspectives in realizing planning and development in the regions.

- ☆ To put in the same level, position, role, responsibility, equality, between men and women as human resources for development.
- ☆ To increase the role and self empowerment of institutions which have roles for women empowerment.

The national strategy also convinces governments and public to create collaboration in gender mainstreaming. There are some examples of such collaboration:

- ☆ In collaboration among the Ministry of Research and Technology, L'Oreal and UNESCO, the Women in Science Programme has beencreated. The programme supports and promotes young women in scientific research in Indonesia through annual selection. Fellowships are awarded to young women scientists at doctoral/postdoctoral level, to undertake a research project in the life sciences field.
- ☆ In collaboration between the ICT Community and the Ministry of Communication and Information Technology, the annual ICT Kartini awards are given to Women who perform well in ICT Industries. They are appreciated for their efforts in inspiring people with their ICT business.
- ☆ The Ministry of Women Empowerment and Child Protection gives annual awards of *Anugerah Parahita Ekapraya* to a person or a group in recognition of their achievement for their efforts in the connection of attaining social welfare through gender equity and equality. The awards are given to the organization leaders or local governments that indicate their roles, commitment and efforts in realizing gender equity and equality throughout PUG strategy.

Women in S&T, especially in ICT, however, are not really promoted yet, although the Ministry of Research and Technology (MRT) and the Ministry of Communication and Information Technology (MCIT) have conducted related programmes to increase research avenues for women scientists and engineers.

Women in Information and Communication Technology (ICT)

The advent of information and communication technology (ICT) provides wide range of possibilities and plays a key role on disseminating information.ICT convergence has risen recently by means of multimedia access, mobile devices convergence and other newly advance technologies. ICT has been progressively developed lately and has become a convenient tool for adopting various requirements. Creative approaches using ICT applications are developed to attract segments of public in particular areas. The popularization of website, blog and social media services are other recent addition to the means of communcation with infinite resources from the internet. Some efforts have been made to take new initiatives in developing various supporting materials.

The Indonesian Government develops programmes by applying ICT for communicating policies, knowledge, and new developments. Figure 2.1 shows the high progress in the increase in Internet users in Indonesia, which might support other S&T programmes by applying ICT. In line with this phenomenon, the

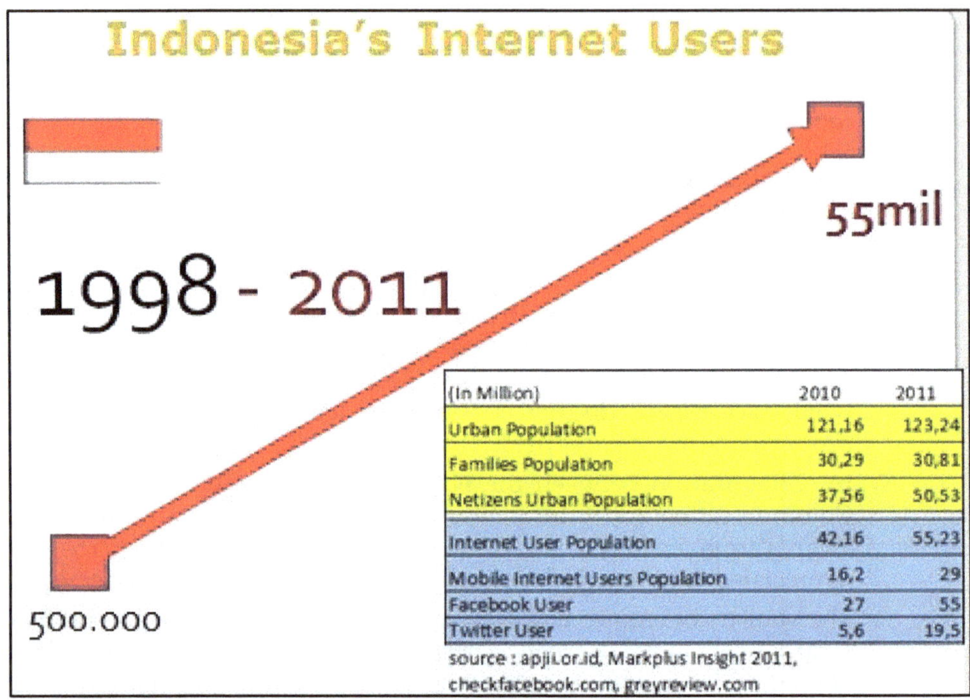

(In Million)	2010	2011
Urban Population	121,16	123,24
Families Population	30,29	30,81
Netizens Urban Population	37,56	50,53
Internet User Population	42,16	55,23
Mobile Internet Users Population	16,2	29
Facebook User	27	55
Twitter User	5,6	19,5

Figure 2.1: Information on the Development of Internet Users in Indonesia.
(*Source*: Ministry of Communication and Information Technology of Indonesia)

Government designs "the Indonesia ICT Roadmap" that bring into reality information to society, with strong ICT – based support.

The programmes developed specially focus on disseminating of S&T and informal learning in the form of utilizing ICT application. Several ICT infrastructure access points at strategicareas within districts and sub districts have been built, in order to provide low cost access as well as information services to public. The example of the programmes are: smart village, mobile tele-center, mobile media center, internet kiosk and creative house.

The fast growth of Social Network Services (SNS) in fact has attracted women to be more familiar with computer and smartphone. According to Inside Facebook report in 2009, there are more than 56 per cent women in US Facebook audience, and amazingly, the most growing segment is the older women (age above 55) with 175 per cent growth rate in 120 days. Unfortunately, this fast growing of computer and gadget users from women segment is not accompanied by their knowledge in ICT. In some cases, this technology illiteracy could bring some issues, such as data security hacked, fraud transaction, etc.

Based on the women empowerment programme and infrastructures of ICT provided at the strategic areas within districts and sub districts, three programmes as case of practices are explained in this paper, to emphasize programme and collaboration. They are: ICT volunteers, Digi Mom, and Batur Geopark Website.

ICT Volunteers

The ICT Volunteers (Relawan TIK) includes collaborative efforts of academics, research and development (R&D) centers, private sectors, government/local governments, and communities. As certain access points are delivered by the MCIT, the access and application should be maintained optimally. Therefore, for continuous operation and implementation, a collaboration should be made by two or more parties, depends upon the nature of core business that has been implemented.

The basic tasks of ICT Volunteers are:

- For education: train about ICT knowledge and application, organize road shows at schools, socialize and train open source applications, training for ICT application at SMEs.
- For partnership: build partnership with central government/local government, ICT organization and communities, companies, that are willing to support ICT Volunteers.
- For socialization and publication: socialize ICT Volunteers in various activities related to ICT, utilize web sites and social media (Facebook and Twitter).

ICT Volunteers can be a mediator in promoting S&T to public: Promoting S&T to public requires interesting and engaging ways as well as creative, attractive and a popular content to cope with the change of a fast-changing world. Besides, the promotion can be applied through the local socio-culture, which is familiar to the public. That can be done through education and media. The scheme is displayed in Figures 2.2 and 2.3.

PROMOTION THROUGH EDUCATION AND MEDIA

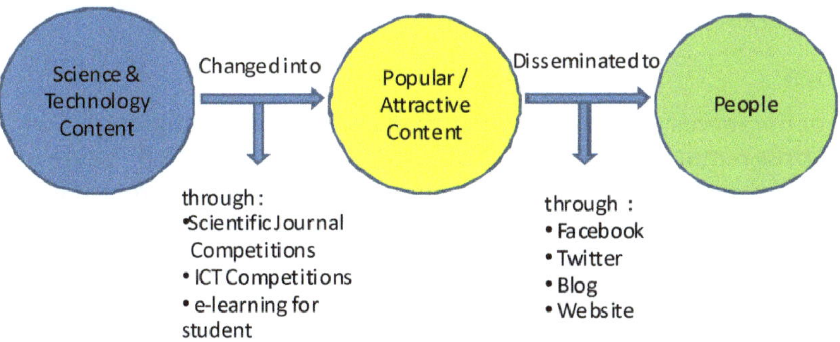

Promoting Science and Technology to the people require interesting and engaging ways as well as creative, attractive and in a popular content to cope with the change of a fast-changing world

Figure 2.2: Scheme of S&T Promotion through Education and Media.

PROMOTION THROUGH CULTURE

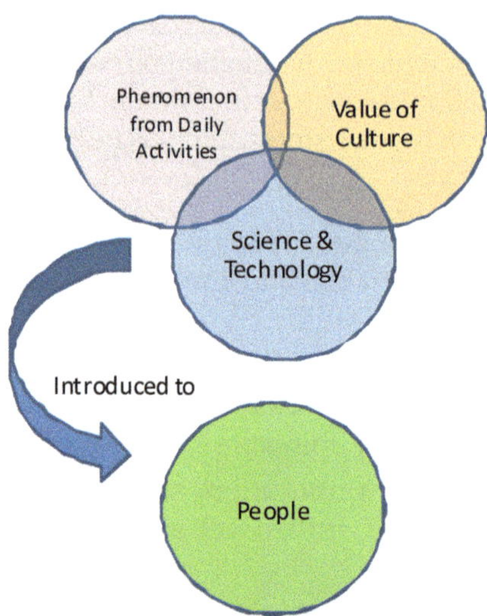

Figure 2.3: Scheme of S&T Promotion through Culture.

The role of women in ICT volunteers is as participants in the education section for socialization and training. The mechanism applied is training of trainers (TOT). Several candidates are selected to be the champions to continue the work and to train the other groups of people. Women were keen in organizing road show at schools to introduce computing and open source software. They are viable in approaching students during the socialization at schools. They would engage with the students playing the computer, while introducing ICT learning to them in attractive or popular contents created by the ICT volunteer team. Once committed, women are dedicated, loyal and can perform equal to men in similar work and training.

DigiMom

One of the success programmes in urban areas is promoting the ICT awareness through a programme called DigiMom. It offers a fun approach by using a blended format of café scientific and workshop. This activity is especially created to empower women through ICT, especially young mothers. The data shows that the fast growth of social network services (SNS) has attracted people especially women to be more familiar with computer and gadget. Figure 2.4 shows the statistics report of Facebook users in Indonesia, where young ages dominating the user profile.

ICT is dominated by men generally, whereas women are more hesitant to learn the advantage of ICT application. In some cases, this technology illiteracy could bring some problems, such as data security mistake, hacked and fraud transaction. To avoid rejection from female audience, the activity is arranged in a café in the form of a fun and leisure workshop. It should be a public place, where women are familiar with the area for their family social activities. The programme is collaborated with famous radio programme for women and science communicator/s.

The workshop starts with ice-breaking session, then active question/answer from enthusiastic audience. The facilitator is also a female engineer in ICT. The participants are asked to set up their account in one or more of SNS, and explained how they can be the active members. Participants are encouraged to share the problems and technical difficulties, as well as creating information in ICT.

There are 2 segments in DigiMom, one is theoritical or technical information regarding digital technologies and its utilization, and the other is to activate or communicatewith their own gadget. The introduction and expert talk consist of: browsing internet effectively, availability of social network services (SNS) in the net and how they could use them for networking and communication, setting up their accounts, or finding work or other resources. The technical information relies on the availability of audience in opening their dimensions, which can be continuously implemented throughout several different segments.

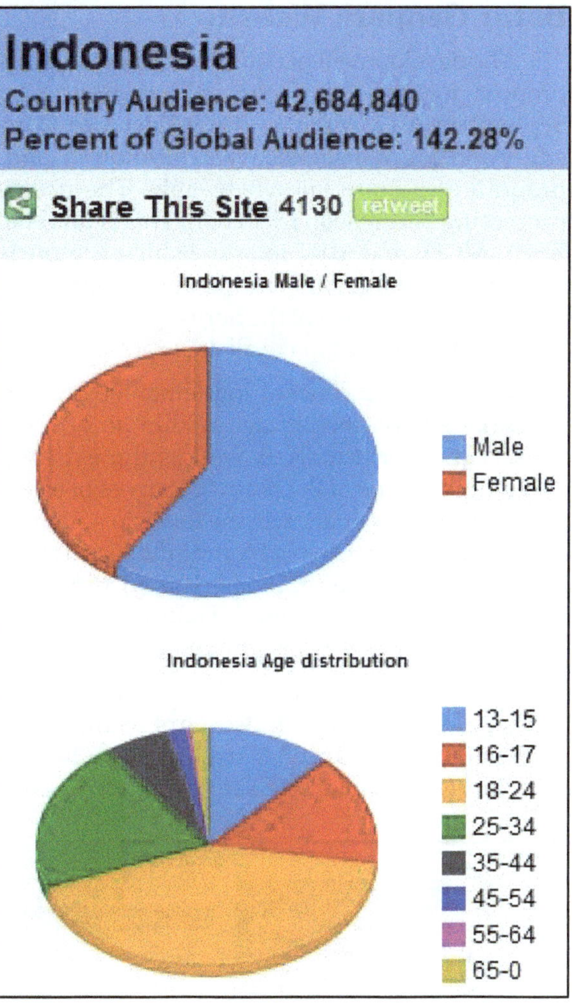

Figure 2.4: Facebook Statistics Report 2012. (*Source*: Facebook Statistic Report 2012)

Quiz and prize is conducted to make the programme lively. A science kid corner is also available with an expert assistance to cater the needs of women accompanied by their children. The activity is continued after the workshop by posting their key points and having open discussion on their Facebook or Twitter accounts. The photographs of the event are asked to be uploaded for sharing. Participants are surveyed to get their opinion and polls, which is both amusing and encouraging, as we are asked to continue the workshop in other themes for the participants.

Batur Geopark Website

The development of Batur Geopark web site is urgently needed, as the Indonesian proposal for the Global Geopark Network was granted by the UNESCO at the end of September 2012. Various disciplines are synergized through informal science education, journalism, festivals, promotion, and these make the implementation of programmes more knowledgeable. The programme is a collaborative effort in promoting Batur Geopark in Bali. This is under a collaboration of local government, BPPT, MCIT, and MRT in establishing ecotourism web site for Bangli District as a geological heritage site. The local government should maintain the establishment of Batur Geopark as one of the heritage sites related to the Global Geopark Network of UNESCO, as well as make all its information available globally.

Geopark is a concept implemented by UNESCO (2000), which is looking at potential site of geoheritage within an area, implemented by the integration of conservation principals as well as the existing planning of that region from the Government. One of the main objectives of establishing Geopark is to increase local community participation in the tourism sector and consequently increases the socio-economic potential of local community, and some engagement with the local people. The other objective of Geopark is to provide sufficient information and public education about the geological nature and importance of certain sites.

The Batur Geopark consists of 25 geosites or geological diversities, explain the history of development of the Batur Volcano complex since 29,300 years ago. It created outstanding geological landscape of scenic beauty. The cultural diversity influenced by the volcano produced many temple buildings (pura) that are made of lava, built with carved volcanic rocks (andesitic and basaltic lava that is produced by the Batur

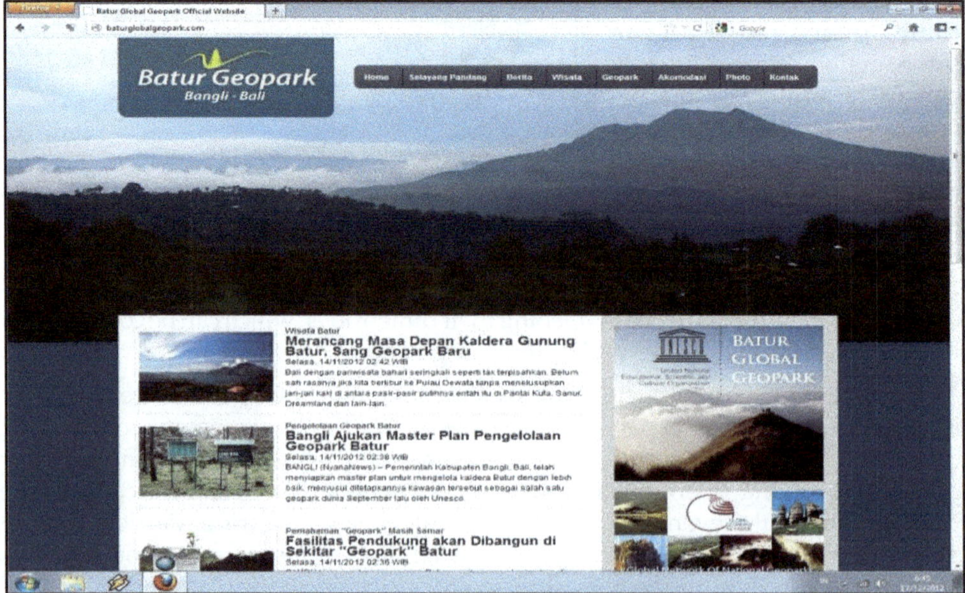

Figure 2.5: Website for Batur Geopark.

Volcano). They are designated as cultural sites and protected as archaeological objects. The Kintamani and Bangli Districts are also famous with good quality of bamboo products that can be developed as building structures and handicrafts for cultural diversity. Thus the Batur Geopark has values for educating public, conservation, and geological heritage for ecotourism, that can be synergized with biodiversity and cultural diversity.

Before this programme was released, the precious information about Batur Geopark was available but scattered at and from various resources, both on-line and off-line. Great efforts have been made for thorough collection of data and information, then to subsequently combine, integrate and synergize the scattered data and information about Batur area and its environment, into a proper information source about Batur Geopark related to ecotourism.

There are two types of training to develop the web site. The first training is to develop the content of the web site of Batur Geopark, where the participants are selected on the basic criteria that they understand fundamentals of IT. They will then train other local people for the continuity of the programme. Role of local women is encouraged to learn the application of input of data for excellent ecotourism. Photographs about the sites were selected by them since they knew better about these places and their popularity

The participants of the second training are mostly women. They are encouraged to undertake the handicraft designs made of bamboo, areas popular for bamboo quality and the local women usually make beautiful handicraft. The products available from other places are more famous although the resources of bamboo come from these area. Therefore, promoting bamboo handicraft from these areas is important, and hence local women were taught to make creative designs of handicrafts made from bamboo.

Women are more meticulous by nature as compared to men, so that they were eager to input their design into the data base in order to be shown in the web site. In the culture of Bali, women are usually hard workers in supporting family's earnings, while men often hang out in the coffee kiosks. Besides, women are more inclined towards education especially for the youth. They are more concerned about the information that should be put in the web site for geological sites. Adding their knowledge in ICT would also make them more independent and self confident. In another way, they earn respect from their children due to increased ICT literacy.

The website was launched during the inauguration of Batur Geopark heritage within the Global Geopark Network of UNESCO, an event which took place at the site in January 2013. The website has then to be maintained and developed further from time to time. The data and information to be displayed in the website must be selected carefully, in order to fulfill the criteria of conservation for geological diversity, cultural diversity and biodiversity, as well as for improvement of education, and socio economic potential.

Summary

ICT media could make S&T public communication more effective as a hub, and stands between scientists and public to provide suitable information through severals channels.The advent of ICT provides wide range of possibilities and plays a key role in disseminating knowledge and information. In a more strategic role, institutional capacity building is required together with the need of champions for S&T communicators. They act as change agent or mediator for the people.

Role of women in S&T should be encouraged, in this sense, since it is necessary to promote S&T culture to youth or students in order to build the future generation of young talented people. Parents who encourage their children to pursue their passions are critically important. Therefore the rise of women performance in public and private sectorsas well as S&T fields must be supported. Their perspectives will add great value to policies and programmes to bring women into all related sectors.

It is, therefore the laws and policies that enforce equality and equity in the workplace are important (sometimes women must face unwritten policies), and for individuals, it should begin from themselves, their goodwill to eliminate barriers that bring women into all sectors, especially ICT.

It is accepted that ICT is attractive to more women, since ICT as the media for S&T communication is not only smart but become urgently needed for nationwide. In case of NAM, there are contemporary S&T-based issues which the governments and people of NAM countries have to deal with, especially empowering women through ICT. People should consider how they can work together to solve these problems, what approaches they should take, and a collective outlook for the future. By using scientific approaches can help to develop and enlarge the capacity to nurture the NAM S&T network.

References

1. Law Number 20/2002 of the National System of Research, Development and Application of Science and Technology, 2002.
2. Legoh, Finarya (2011); Overview Women in Science and Technology in Indonesia, presented in the 2011 Asia Women Eco-Science Forum (2011 AWESF) held at Seoul, 7-9 November 2011.
3. Legoh, Finarya;Permatasari, Dyah Ratna (2012); Café Scientifique and Workshop to Empower Women in Information Technology presented in the PCST 2012 International Conference (Public Communication on S&T) held at Florence, 18-20 April 2012.
4. Legoh, Finarya (2012); Practice and Future Science and Technology Communication in Indonesia, presented in the International Forum on Public Communication of S&T Studies held atBeijing, 17-19 August 2012.
5. Legoh, Finarya; Djamsari, Mustadjab; Sadoso, Hendro (2012); The application of ICT in Increasing the Potential Tourism Promotion of Bali; a project activity report of PKPP 2012 submitted to the Ministry of Research and Technology, October 2012.

6. Legoh, Finarya (2013); Promoting Public Communication on Science and Technology through ICT, presented as a Country Report in the S&T Communication Leadership Workshop II of ASIA Forum held atPathumThani, Thailand, 3-5 April 2013.
7. www.menegpp.go.id
8. id.wikipedia.org/wiki/Kongres_Wanita_Indonesiaý
9. Facebook statistics reports

Chapter 3

The Roles and Responsibilities of Academic Institutions to Encourage Women Participation in ICT Related Disciplines

Shireen K. Assem

Deputy Director, Professor of Biotechnology,
Agricultural Genetic Engineering Research Institute (AGERI),
Agricultural Research Center (ARC), Cairo, Egypt
E-mail: shireen_assem@yahoo.com

ABSTRACT

Information and communication technologies (ICT) play a significant role in the socioeconomic development of any country and it is a key contributor to productivity and competitiveness. Historically, the isolation of women from the mainstream economy and their lack of access to information are because of societal, cultural and market constraints. These have led women to become distant from the global pool of information and knowledge. Information and communication technology sector appears to be a promising field for improving women engagement in labor market and the community, however, more efforts should be devoted to increase their engagement. To be able to reap from the benefits of ICT, women must be equipped with skills to prepare them for a range of roles, not only as ICT users, but also as creators and designers. To meet continuing growth in ICT more women should consider ICT related careers. The key element in attracting women to ICT is education. However, the difficulty to attract and retain female students in ICT has been well recognized in several countries. Science and technology are intimately connected with development, because they are key ingredients of solutions to the most serious poverty alleviation and economic development challenges that we currently face integrating the activities with academic institutions to encourage women participation in ICT related disciplines, and subsequently and are likely to face in the future. Capacity building and technology transfer are integral parts of the process of empowering women in developing countries through ICT. Academic institutions in different countries should create appropriate institutional framework to allow and encourage women to pursue a professional career in ICT while respecting possible constraints to their professional choices. This research paper discusses the possible opportunities for academic institutions in empowering women through ICT and the responsibilities of different sectors in reflect on women overall development.

Keywords: *ICT, Empowering women, Academic institutions, Gender equity, Girls' education, Development.*

Introduction

Globalization has brought in positive prospects for helping economies to be transformed to knowledge based economies. A knowledge based economy is defined to be one which can efficiently create, disseminate, and use knowledge to enhance its growth and competitiveness. Being a main catalyst for economic development, knowledge, can significantly contribute in raising the living standards of nations and individuals, by narrowing different kinds of gaps (income, gender, education, etc.), (Mandour, 2009).

Information and Communication Technologies (ICT) are a diverse set of technological tools and resources to create, disseminate, store, bring value-addition and manage information, (Nath, 2001). ICT has greatly changed our lives. It plays a significant role in the socioeconomic development of any country and it is a key contributor to productivity and competitiveness. It represents one major pillar in forming a knowledge based economy through facilitating effective communication, disseminating, and processing of information, (Mandour, 2009). ICT as an industry (hardware, software and services) has not only provided means to help generate, store, process and communicate information globally, but has also pervaded all sectors of the economy. They act as integrating and enabling technologies in order to improve productivity for sustainable development (Sunday and Oludele, 2010). Knowledge networking opens up a new way of interactive communication between government bodies, NGOs, academic and research institutions, and civil society. It helps communities, both men and women, to take appropriate steps to recognize and document the knowledge they possess, and in reflecting this knowledge in a wider social domain for directed change through the use of information and communication technologies (Nath, 2001). With the growing use of various forms of ICT, it is becoming a priority to utilize such technologies in serving socioeconomic developmental objectives (Kamel, 2007).

ICT has been an integral part of development. It is now strongly recognized as an all-purpose enabling tool for development in any country. Sunday and Oludele (2010) summarized the Millennium Development Goals in few points: 1. Eradicate poverty and hunger, 2. Achieve universal primary education, 3. Promote gender equality and empower women, 4. Reduce child mortality, 5. Improve maternal health 6. Combat HIV/AIDS, malaria, and other diseases, 7. Ensure environmental sustainability, 8. Develop a global partnership for development. Within these goals, empowering women is the most effective one that can help achieving most of the other goals as women are an effective power and equal partner in any society. Sunday and Oludele (2010) concluded that it is now a well understood fact that without a progress towards the empowerment of women, any attempt to raise the quality of lives of people in developing countries would be incomplete.

Women and Education

Gender inequality tends to slow economic growth, according to Christiaan Poortman, World Bank Vice President for the Middle East and North Africa: *"No country can raise the standard of living and improve the well-being of its people without the participation of half its population"*. Experience in other countries have shown over and over again that women are important actors in development, to hold them back is to hold back the potential for economic growth.(World Bank, 2004, online, Mbarika *et al.*, 2007).

Education is actually a key means of empowering women and is in itself a human right. Educated girls lead better lives. Education is especially central to women's empowerment in so far as it enables women to become more productive both inside and outside the household. Investments in women's general education, including literacy is considered one of the most important elements, complementary to income-generating activities that are considered essential for women's economic empowerment. Post-primary education has the greatest payoff for women's empowerment in that it increases both income earning opportunities and decision making autonomy (Abdel Mowla, 2009).

The key element in attracting women to ICT is education. Girls have to receive good education in the primary education years with introduction to ICT information and tools. This can be further continued successfully by attracting girls to ICT studies and applications during their high school years. However, the difficulty to attract and retain female students in ICT has been well recognized in several countries. The low number of female students studying ICT in secondary school tends to decrease further at tertiary level, (Miliszewska and Moore, 2010).

A report compiled by the European Schoolnet and CISCO Systems Inc. indicated that, out of female students across Britain, France, Italy, Poland, and the Netherlands, who like studying ICT at high school, most fail to progress those studies to tertiary level or translate their computer competencies into ICT careers (Gras-Velazquez *et al.*, 2009).

Education during the school years plays a vital role in attracting both girls and boys to ICT studies and career. The most important elements in attracting girls to ICT is simplify the information, such as provide computers and softwares that attract their attention and make ICT lessons more enjoyable than just information to learn for the exams. Later at high school, girls could be encouraged by practicing ICT in different applied projects related to their lives.

In developing countries, limited funding resources could be one of the constraints in promoting ICT education process, were the government resources are very limited. This could be enhanced by contribution of various sectors of the society including the private sector in funding educational endeavours at schools. The funding could be for necessary tools and facilities that support ICT study at different educational stages. Maintaining a whole some process of education then should be monitored by the government, to ensure equality and perfection in the complete process. Another limiting factor is the knowledge of teachers involved in ICT education process. School teachers engaged in the ICT teaching should receive special support from the government to attain a minimum standard level of ICT teaching. They should not only be well qualified to ICT education but also have the ability to better communicate with young generations by simplifying the information and guiding them during the ICT lessons.

A number of studies argued that ICT has a positive impact on gender gap through facilitating the engagement of women in the labor market and its contribution to women's empowerment and capacity building in numerous ways (Maier and Nair, 2008). Since ICT related activities require essential intellectual abilities rather than physical abilities, and also allow for more flexible working conditions, women could enjoy comparative advantage when compared to men. Moreover, ICT has facilitated the transfer of information and knowledge around the world which enable women to acquire education through distance learning, and hence assist them in acquiring education in the fields that they were previously deprived from. Thus, the relationship between ICT and education is a two-way relationship, where education is arguably the most important prerequisite that enables women to take advantage of the opportunities offered by ICT, and ICT per se facilitates education attainment (Chen, 2004). Moreover, as Mitter (2006) underpinned, women entrepreneurs can benefit from better access to global information and markets which would enhance their competitiveness in the market (Mandour, 2009).

Education Level and Quality

The effect of education on women's economic empowerment is determined mainly by the level and quality of education that they have received. Education quality is determinate for women's economic empowerment. Education quality indirectly affects women's economic participation and opportunity through affecting the likelihood of enrolment and the education level, (Abdel Mowla, 2009). Empirical evidence suggests that standard aspects of school quality have a stronger impact on girls' education than on boys' education. On one hand, evidence suggests that quality is an important demand factor. When education costs too much and when good quality education is hard to come by, parents, especially those in under- privileged groups, may feel that the future returns may not justify the present costs. Moreover, girls' enrollment is more sensitive than that of boys to school quality. On the other hand, access to high-quality education increases the likelihood of achieving higher levels of education. It influences earnings by affecting college-choice behavior. High quality high schools increase the probability of college attendance, which generates a persistent wage gain relative to wages that an individual could expect to earn with no college education (Task Force on Education 2005; Strayer 2002). In addition, education quality directly affects both women's economic participation and opportunity through affecting skills they acquire.

The Importance of ICT for Women Empowerment

ICT is for everyone and women have to be an equal beneficiary to the advantages offered by the technology, and the products and processes which emerge from their use. Throughout generations, women have been isolated from the mainstream economy with lack of access to information because of societal, cultural and market constraints. This has led them to become distant from the global pool of information and communication technologies. This distance is reflected in the levels of empowerment and equality of women in comparison to men, and has enormously contributed to the slow pace of development.

It is now a well understood fact that without progress towards the empowerment of women, any attempt to raise the quality of lives of people in developing countries would be incomplete. There is an increasing amount of evidence which substantiates that societies that discriminate by gender pay a high price in terms of their ability to develop and to reduce poverty (Nath, 2001).

Globally, women's access to information has been a major concern. The United Nations places lack of access to information as the third most important issues facing women after poverty and violence (UNESCO, 2003). Gras-Velazquez *et al.*,(2009) unsurprisingly found a direct correlation between the enthusiasm of mothers for ICT and that evidenced by their daughters. Most of mothers surveyed are part of the European Schoolnet-CISCO project. They were pleasingly positive about ICT studies. More worrying, however, was the correlation that existed in perceptions about ICT careers, with female role models and the girls they influence believing that jobs in ICT were less likely than other careers to involve travel, helping others, or working independently.

Education Institutions and Research Programmes

Equal access to education and equal opportunity in gaining the skills are necessary for women to compete in the labor market. The better educated a woman is, the more able and willing she is to compete with men in the labor market. Gains in women's education lead to increases in their productivity. This in turn reduces discrimination against them. This is obviously evident in today's labor markets, where jobs are becoming more and more demanding of skills. As a result, workers need to upgrade their skills or else they will be at a risk for losing out in the competition for jobs. This is the reason why many of the unemployed might be considered as "unemployable in a modern economy" due to their comparatively low level of education.

McCullough (2002) has identified a number of programmes across the United States of America. The educational establishments have worked either by themselves or in collaboration with community groups, public institutions, or government agencies, to raise female participation in technology related fields of engineering, computer science, mathematics, and physics. As McCullough (2002) suggested, these programmes, which are often funded through the National Science Foundation (NSF) Gender Equity Programme, vary in size. These programmes often call for a strong collaboration with local industries, reach out to parents, and frequently involve residential workshops of up to ten days duration. In addition, girls identified as being disadvantaged gateways into science and technology careers are provided with greater access to education in science, mathematics, engineering, and technology, and programmes such as SECME3 RISE (Raising Interest in Science and Engineering). The teachers are also provided with professional development opportunities around gender equity and the integration of technology into school curricula (McCullough, 2002). In spreading themselves, thus, many of the programmes have attempted to satisfy various theoretical perspectives found in the literature, such as: the importance of parents as role models, the age at which intervention must occur, the need to raise gender awareness in the classroom and the school, and so on (Miliszewska and Moore, 2010).

Science and Technology in Relation to Women and ICT

When women are compared with men, especially in developing countries, in general women have fewer years of schooling, lower enrollment rates in science and technology education, and less work experience. In addition, they often lack access to skills training that would enable them to gain specialized in ICT related jobs. Thus, with respect to men, women's wages are relatively lower if a premium is dedicated to human capital. Another major constraint that hinders women from benefiting from ICT is the fact that the majority of the newly created technology and telecommunication jobs in developing countries are in the private sector. In this area the number of women-specific related benefits remains small when compared to public sector. As a result, women more often opt for public sector work that is more likely to offer childcare provision, flexible hours and maternity leave, but lagging behind in ICT related aspects (Mandour, 2009).

Technologies are socially constructed and thus have different impacts on women and men. Women's capacity to exploit the potential new ICT as tools for empowerment is constrained in different ways. Some constraints are linked to factors that affect both women and men, including technical infrastructure, connection costs, computer literacy and language skills. These overall constraints are, however, exacerbated in many cases by gender-based determinants which particularly pose as real disadvantages for women (Women 2000 and beyond, Sept. 2005).

However, the proportion of women in industrial and academic science has shot up over the past 20 years. According to the US National Science Foundation, women make up 25 per cent of tenured academics in science and engineering and more than 25 per cent of industry scientists in research and development. But when it comes to academics engaging in commercial work, patenting their discoveries, starting biotech companies or serving on SABs, the numbers are less progressive (McCook, 2013).

ICT is linked to science and technology in the development process. ICT is one way of facilitating and ensuring that the current and the future generation of young people, especially young girls are more interested in science and technology. Science and technology also play a big role in the improvement of broadband infrastructure. Without vital infrastructure, ICT cannot be utilized to its fullest potential. Moreover, lots of ICT development takes place in developing countries like Kenya, Egypt and South Africa (in Africa) and others in Asia and Europe. Without good infrastructure, indigenous research aided by investment in science and technology and empowering women in these fields will be difficult in reaching development goals (Taylor, 2010).

Capacity Building and Technology Transfer

There is a need for capacity building. Civil society and foundations should play a big role in developing capacity in this area. This can be done by facilitating the exploration of links between ICT and the development of science and technology. Partnerships across countries, regions, academic institutions and disciplines are also the key ingredients in the success of many initiatives.

The appropriate use of ICT could strengthen and increase the capacity of women's organizations. The facilities can be increasing women's access to ICT, provide network opportunities, increase capacity of ICT skills for women and girls, raise awareness on the benefits of ICT, and train women's organizations in using communication technologies in their information strategies. The role of the organizations of empowering women through ICT is activated by increasing awareness of girls and women of their existing potentials and skills and help to deepen them, explore and develop new and relevant applications to meet specific needs, share and exchange information among them.

To sustain these capacities, many projects have been developed in several countries for training women's organizations in the use of knowledge sharing strategies in their research, planning, advocacy, information collection, dissemination and communication strategies. ENAWA Project of "Women's Information Technology Transfer" (WiTT), Stockholm Challenge (2010), is one example of such projects aiming to empower women organizations through ICT.

In ENAWA project, ICT contributed to the organizational objectives, and the outcome of this project concludes that women were mostly inspired. Women are not only ignorant in respect of ICTs but also need a great boost in self-confidence and motivation to apply themselves and have skills in all areas of life. Moreover, a minority of women have been able to apply and get better paying jobs, with an increase in salary. When evaluated, most of the women trained feel "happy" as they are able to communicate and share information, which is valuable to their life and life improvements. It has also increased girl student's educational skills at school.

Conclusions

Different sectors are responsible for integrating with academic institutions to encourage women participation in ICT related disciplines. The development should promote science and technology and ICT simultaneously. Governments, academics, IGOs, civil society, scientists (men and women) and all other stakeholders need to be armed with tools and skills to help them share and build scientific and technological knowledge and capacity for achieving sustainable development.

References

1. Abdel Mowla S.A.A. (2009). Education and Economic Empowerment of Women in Egypt, SRC/CIDA Research Programme on Gender and Work, Working Papers Series, *Working Paper #002.*

2. Alison McCook (2013). Women in Science, The Gender Gap and How to Close it. Nature Special: nature.com/women.
3. Chen Derek (2004). "Gender Equality and Economic Development, The Role of Information and Communication Technologies", World Bank Policy Research Working Paper No. 3285, April 2004.
4. Gras-Velazquez *et al.* (2009). Women and ICT: Why are girls still not attracted to ICT studies and careers? White paper, Insight observatory for new technologies and education, European SchoolNet. Retrieved 5 November, 2009 from http://insight.eun.org/ww/en/pub/insight/misc/specialreports/eun_white_paper_women_it.htm
5. Kamel, S. (2007). *The Evolution of ICT infrastructure in Egypt in Science, Technology and Sustainability in MENA*, Inderscience Enterprises Ltd.
6. Maier S. and Nair-R. U. (2008). Empowering Women Through ICT-Based Business Initiatives: An Overview of Best Practices in E-Commerce/E-Retailing Projects, Information Technology and International Development Journal, The MIT Press 2008, Volume 4, Number 2, Winter 2007, available at: http://www.mitpressjournals.org/doi/pdf/10.1162/itid.2008.00007.
7. Mandour D. A.Impact of ICT on gender gap in Egypt (2009). SRC/CIDA Research Programme on Gender and Work, Working Papers Series, *Working Paper #004.*
8. Mbarika V. W. A., Payton F. C., Kvasny L. and Amadi A. (2007).IT Education and Workforce Participation: A New Era for Women in Kenya? *The Information Society*, 23: 1–18.
9. McCullough, C. (2002). Attracting under-represented groups to engineering and computer science. Proceedings of the American Society of Engineering Education Southeastern Section Conference (pp. 1- 11), Gainesville, Florida. Retrieved 5 September, 2009 from http://155.225.14.146/aseese/ proceedings/ASEE2002/P2002046COEETMCC.pdf
10. Miliszewska I. and Moore A. (2010).Encouraging Girls to Consider a Career in ICT: A Review of Strategies. *Journal of Information Technology Education*, Volume 9: Innovations in Practice, Editor: Lorraine Staehr.
11. Mitter S. (2006). Globalization and ICT: Employment Opportunities for Women, Part III, available at: http://gab.wigsat.org/partIII.pdf.
12. Nath V. (2001). Empowerment and Governance through Information and Communication Technologies: Women's Perspective. ntl. Inform. and Libr.Rev. 33, 317^339. Available online at http://www.idealibrary.com.
13. Stockholm Challenge, (2010).http://www.stockholmchallenge.org/project/data/witt-womens-information-technology-transfer.
14. Strayer, Wayne (2002). "The Returns to School Quality: College Choice and Earnings". *Journal of Labor Economics.* Vol. 20, No. 3, pp. 475-503.
15. Sunday A. and Oludele A. (2010). Information and Communication Technology (ICT) Revolution: Its Environmental Impact and Sustainable Development. *International Journal on Computer Science and Engineering*, 2 (01S): 30-35.

16. Task Force on Education and Gender Equality (2005a). Toward Universal Primary Education: Investments, Incentives and Institutions. London.
17. Taylor K. (2010) The link between ICT and Science and Technology in Africa: Implications for Civil Society Organisations. GuideStar International's Blog at http://blog.guidestarinternational.org/

Chapter 4

Empowering Rural Women for Socio-economic Development through Information, Communication and Technology: Challenges and Opportunities for Malawi

Catherine Chaweza

Chief Information Officer,
National Commission for Science and Technology,
Lilongwe 3, Malawi
E-mail: katechaweza@gmail.com; katechaweza@yahoo.com

ABSTRACT

Rural women in Malawi are generally the fiber of society and are vital for the wellbeing and progress of communities. They have responsibility to raise children and reinforce communities. This also includes being responsible for their children's education needs, providing nutritional needs of families and, more often than not, women are breadwinners in their families and are deeply involved in agricultural activities. At the community level, the Malawian woman is perceived as the caregiver on whom the sole responsibility to care for sick family and relatives rests. Most of the rural women go about this responsibility with no advanced information whatsoever. Therefore, their lives have remained ordinary and mediocre in the absence of modern information to empower them for the improvement of the quality of their lives and their communities. The use of information and communication technology (ICT) by these rural women would therefore, go a long way to empower them, to move from the subsistence level to a point where they could make a positive and meaningful contribution to socio-economic development of the country.

Currently, the access to information and communication technology (ICT) by rural women in Malawi is almost at zero, except for the use of telecommunication, more specifically the cellular phone. Quite a considerable number of rural women use the cellular phone just for simple procedure of receiving and making telephone calls. The challenges that have obstructed the access and use of ICT by the rural Malawian woman are many. The ICT policy highlights one of the greatest challenges as the low level of education resulting in high illiteracy rate that makes it difficult to implement ICT programmes particularly amongst women, youth, elderly and other disadvantaged groups. Apart from that, other challenges include lack of ICT infrastructure, expensive internet facilities and lack of basic training to enable use of ICT. Another pertinent challenge is underdeveloped telecommunication and postal infrastructures in the rural areas, plus the perceived lack of relevant internet content that may be useful to the rural population. This paper makes recommendations of how the existing challenges can be addressed.

The new ICT policy for Malawi has a deliberate gender agenda. The policy takes particular interest in the involvement of women to access and use ICT, with a special focus on rural women. The opportunities discussed in this paper include a setup that has already been introduced in Malawi and is called the Telecenters. The Government of Malawi has established telecenters in most rural trading centers of the country.

The paper discusses how these telecenters would be used as a hub for information transfer to the rural women through ICT. The information would be relevant to their situation and particular needs. The paper also highlights how the Public Private Partnership that the Government of Malawi is encouraging, at the moment, could enhance the access and use of ICT by rural women through the coordination between Government Departments, private sector, NGOs and relevant Community Based Organizations. The policy also encourages the training of women in ICT. One of the strategies for implementation of the policy states that, "the Ministry of Women together with the Ministry of Local Government and the Ministry of Finance shall take the lead to support and facilitate the training of women, youth and the disadvantaged in key skills required by the information and knowledge economy".

The paper, therefore, discusses how rural women would be supported to access and use internet for the improvement of quality of life. It will also highlight on how the use of computer and internet can empower rural women politically, socially and economically. It also suggests ways through which the Government of Malawi, local NGOs, private sector and government agents would ensure that the grass root Malawian women benefit from ICT for overall socio-economic development of the country.

Keywords: *Access and use, Empowerment, Socio-economic development, Community, Benefits, Support.*

Profile of Malawi

The Republic of Malawi is a landlocked country in Southeast Africa. It is bordered by Zambia to the northwest, Tanzania to the northeast, and Mozambique on the east, south and west. The country is separated from Tanzania and Mozambique by Lake Malawi. Malawi is over 118,000 km^2 (45,560 sq miles). Its capital is Lilongwe, which is also Malawi's largest city. The country is also nicknamed "The Warm Heart of Africa". Malawi has over 15,000,000 inhabitants with over 52 per cent of the population being women. Malawi has a literacy rate of 71 per cent and women are the most illiterate.

Malawi is among the world's least-developed countries. The economy is heavily based on agriculture, with a largely rural population. The rural Malawi population is estimated at 80 per cent. The Malawian government depends heavily on outside aid to meet development needs, although this need (and the aid offered) has decreased since 2000. The Malawian government faces challenges in building and expanding the economy, improving education, health care, environment, and becoming financially independent. Malawi has several programmes developed since 2005 that focus on these issues, and the country's development indicators have been improving. There have been improvements in economic growth, education and healthcare seen around 2007 and 2008. However, recently Malawi has experienced a huge economic reduction with its currency, the Malawi Kwacha, devaluing drastically and falling against the dollar. In one year the Kwacha moved from 130 to over 400 Kwacha to the dollar. This depreciation in the value of Kwacha translated into lack of crucial necessities and services, for instance, the lack of necessary drugs in the country's hospitals and there have been serious fuel shortages and food scarcity resulting in lack of necessary services. The prices of basic necessities, including food have also suddenly risen.

Almost all the rural Malawians are subsistence farmers, in order to feed their families and selling very little of their produce mostly, at the beginning of the harvest season. Women in Malawi are heavily involved in the growing of crops and are the care givers of the sick in their societies. There are a lot of women headed homes which give extra responsibility to women to cater for their homes. Most rural women are illiterate but studies have indicated that women are the most responsive to adult learning. In cases where adult literacy has been introduced in Malawi, women have constituted over 90 per cent of the learners.

Most Malawian women are involved in small and medium enterprises as a source of income for their homes. The businesses range from selling fritters, fruits, vegetables and bigger businesses involve rearing chickens to sell meat or eggs. The more advanced business women own small grocery shops, mainly out of small shacks built near their homes. There is also a cross border trade which very few rural women are involved in. Other than that, most women depend on piece work. Others get involved in the Malawi Government "food for work" programme where they work for a small pay at the end of every week. Most rural women are beneficiaries of the Malawi Government Farm Input Subsidy Programme (FISP), where they are given coupons that they use to acquire cheap farm inputs and fertilizer, to ensure their families are food secure.

Status of ICT in Malawi

The Malawi Growth and Development Strategy (MGDS) is the overarching medium term strategy for Malawi designed, to attain Malawi's long term aspiration as outlined in its Vision 2020. The strategy covers a period of six years of 2011- 2016.

Among the priority areas of the Strategy, it outlines infrastructure development including the development of ICT.

Information and Communication Technologies (ICT) and services importantly contribute to the development of the country. As outlined in the Malawi Growth and Development Strategy (MGDS II), "information is a vital resource that should be made available in a form that is applicable and usable, and at the right time for the citizens to make informed decisions."

The Government of Malawi realizes the role of ICT sector in the development of the country and has put in place a number of mechanisms to further this end. The Government therefore makes commendable efforts towards the development of ICT. In the period of 2006-2011 during the implementation of MGDS 1, the ICT sector made commendable achievements by the connection of fiber optic cable. This has led to the improvement of the delivery of telecommunication services and improved mobile and telecommunication coverage.

The goals of the ICT sector in Malawi are:

- To increase utilization of ICT
- To ensure universal access to ICT products and services, to improve service delivery in both public and private sectors and
- To ensure that the population has access to timely and relevant information, and increase popular participation of citizens in development, governance and democratic processes.

Despite a number of challenges, the Government ensures better access to information by using a number of strategies, such as: developing reliable, fast and robust national ICT infrastructure that feeds into international networks, improving efficiency in delivering postal services and developing public online services. It also envisages improving the ICT broadband infrastructure, increasing usage and access to information and communication services, improving postal and broadcasting services, improving ICT governance and enhancing ICT capacity for the general public.

The Government has also developed the ICT policy which has just been adopted. The policy aims to further develop in the ICT sector and to promote the integration of ICT in all sectors of the country's economy. The policy also has a specific gender agenda.

The Government department that is responsible for the development of Information, Communication Technologies is the Ministry of Information and Civic Education (MICE). This ministry oversees overall sectors to provide the enabling environment of ICT development. This ministry is also responsible for the development and implementation of the ICT policy.

The ICT sector in Malawi is regulated by a number of Policy and Legal instruments namely, The Communications Sector Policy Statement (1998) and the Communications Act of 1998. The Communications Sector Policy was formulated by the Government of Malawi through the Ministry of Information and Civic Education

(MICE), while the Communications Act (1998) was formulated by the government through MICE and the Ministry of Justice by consulting to the general public.

The Malawi Communications Regulatory Authority (MACRA) is the government body which is mandated to regulate fair competition through leveling the playing field and to ensure the provision of reliable and affordable communication services. Operators are responsible for building, maintaining info-communication infrastructure, providing efficient and affordable ICT services to Malawians (rural and urban). MACRA was established under section 3 of the Communications Act (1998). It is mandated to regulate the provision of:

- ☆ Telecommunication networks and services
- ☆ Broadcasting
- ☆ Postal services and
- ☆ Management of radio frequencies (spectrum)

MACRA regulates the communication sector by, among other things, doing the following:

- ☆ Ensuring the provision of reliable and affordable communication services.
- ☆ Protecting the interests of consumers.
- ☆ Promoting efficiency and competition.
- ☆ Encouraging the introduction of new communication services.
- ☆ Promoting research and development.
- ☆ Fostering the development of communication services and technology in accordance with recognized international standards.
- ☆ Promoting universal access of ICT services.
- ☆ Regulating fair access to scarce and vital resources *e.g.* radio spectrum.
- ☆ Serving as a policy advisory body to the Ministry responsible for communication.

There are a number of ICT Development Projects in the country which include "the connect a school" and "the connect a community project". The 'connect a community' project involves providing internet connectivity to rural communities, while the connect a school aims to provide connectivity to most schools in the country. These projects are done through the following initiatives:

- ☆ *The UA Project*: the project is aimed to provide internet connectivity and public payphones to ten districts in the country, which is funded by the World Bank and the Government.
- ☆ *The ICTs for Sustainable Rural Development Project (ICTSRDP)*: it establishes internet connectivity to rural areas of two selected districts in the country.
- ☆ *ITU/MACRA/MPC Projects*: they establish internet connectivity to some district centers and postal offices of the country.

- *Internet Connectivity to Universities*: there is also a project to enhance the ICT capacity of universities and it has been done for two universities so far. The Government plans to connect more schools and colleges through other initiatives.

The Government of Malawi is also implementing what is called the Regional Communications Infrastructure Project-Malawi (RCIPMW) with a grant obtained from the World Bank. The project is being implemented through a Public Private Partnership (PPP) model following pro-competition and open access principles. There are a number of institutions which are implementing the project namely the Privatization Commission (PC) in cooperation with the Ministry of Information and Civic Education (MICE) and MACRA. The project aims to connect Malawi to the high speed submarine cables by providing an inland virtual landing station for these submarine cables from Mozambique/Tanzania.

Challenges

The ICT sector in Malawi is facing a number of challenges as follows:

- High cost of telecommunication services in the country
- Unavailability of ICT infrastructure in most rural parts of the country
- Electricity power failure
- Penetration of computers concentrated only in private and public sectors
- Limited computer penetration in homes
- Over 90 per cent of Malawians have internet access only at the office
- Almost all ICT equipment is imported which makes them expensive
- Lack of a National ICT Policy
- Inadequate legal and regulatory frameworks
- Inadequate Institutional and Human Resource Capacity
- Under developed ICT Infrastructure
- Absence of data on some ICT Indicators.

Opportunities

Regardless of the challenges that the ICT sector is facing in Malawi, there are a number of opportunities that indicate that in the near future there will be an improvement in the sector. The opportunities are as follows:

- There is Government political will
- Generous tax allowance on telecommunications equipment
- ICT Policy and other regulating frameworks and policies
- Skilled labour market: ICT has been included in the syllabus
- Network of stakeholders: there is constant consultation on national issues that pertain to the ICT sector

Programmes and Interventions to Address the Challenges

The Government has come up with a number of programmes aimed to address the challenges outlined above:

- Developing a reliable, fast, adaptive and robust national ICT infrastructure that feeds into international networks
- Enhancement of the National College of Information Technology and the introduction of electronic learning facilities in order to increase the number of ICT skills in the country
- Establishment of a national Data Centre that will rationalize data storage and data processing facilities
- Introduction of at least five electronic systems to facilitate efficiency and effectiveness in public service delivery
- Development and implementation of a web portal that will facilitate online communication between Government and the general public
- Mainstreaming ICT into sector of policies and strategies and operations
- Improving ICT access service to rural and underserved communities
- Development of e-government to aim for a paperless way of doing Government communication
- All Government ministries and departments are connected through the Government Wide Area Network (GWAN)
- Developing a national ICT Indicator database.

Telecentres in Malawi

The telecenter project in Malawi has largely been implemented through MACRA in fulfillment of its functions "to promote universal access to ICT services". Telecenters are public places where ICT can be accessed by individuals and communities for improvement and development. These places can be described generically as Multi-Purpose Community Telecenters (MCTs), Public ICT Access Points (PIAPs), etc. Much of the literature emphasizes to ensure the Telecenters should meet the information and communication needs of the community. Telecenters offer a wide range of communication services to public and well utilized Telecenters could improve social and economic welfare of rural communities.

The first telecenter in Malawi was operated in 2007 at a place called Goliati in the southern part of the country. Malawi has five models of telecenters as described below:

1. Local management committee (LMC)
2. Entrepreneurship model- run by entrepreneur
3. Postal telecenters run by Malawi Postal Corporation (MPC)
4. Operator managed telecenters
5. School/College telecenters

Table 4.1: Status of ICT Indicators in 2010/2011 in Relation to Previous Years

Performance Indicator	*Proposed Definition*	*Measuring Unit*	*Baseline 2010*	*Target 2011*	*Actual 2011*	*Data Sources/ MoV*	*Approach/ Metho-dology of Data Collection*	*Schedule/ Frequency*	*Person/ Entity Responsible*
Number of Computers per 1000 population	Number of computers installed per 1000 population. The statistics include PCs, laptops, notebooks etc, but excludes terminals connected to mainframe and mini-computers that are primarily intended for shared use, and devices such as smart-phones and personal digital assistants (PDAs).		–		–	Report	Administ-rative Data	Annually	E-Govern-ment
No. of Fixed Phone Subscribers per 100 population (Fixed Line Density)	Number of fixed telephone lines per 100 population	Number	0.85		1.19	Report	Administ-rative Data	Annually	MACRA
No. of Mobile Phone Subscribers per 100 population (Mobile Density)	Mobile cellular telephone subscriptions per 100 population	Number	16.9		17.13	Report	Administ-rative Data	Annually	MACRA
No. of Internet Service Providers	Count of Internet Service Providers	Number	10		12	Report	Administ-rative Data	Annually	MACRA
Type of Internet Connectivity						Report	Administ-rative Data	Annually	MACRA
1. Broadband Internet subscribers	Number of entities or individuals who pay for high-speed access to the Internet (a TCP/IP connection). High speed access is defined as being equal to, or greater than 256 kbit/s	Number	86,957		473,684	Report	Administ-rative Data	Annually	MACRA

Contd...

Table 4.1–Contd...

Performance Indicator	Proposed Definition	Measuring Unit	Baseline 2010	Target 2011	Actual 2011	Data Sources/ MoV	Approach/ Metho-dology of Data Collection	Schedule/ Frequency	Person/ Entity Responsible
2. ADSL		Number	8,005		9800	Report	Administ-rative Data	Annually	MACRA
No. of Internet Users per 100 population	Total number of users of internet services	Number	–		–	Report	Administ-rative Data	Annually	MACRA
No. of Internet subscribers per 100 population	Number of entities or individuals who pay for access to Internet	Number	2.26		3.33	Report	Administ-rative Data	Annually	MACRA
Percentage of population covered by mobile cellular telephony	Percentage of a country's inhabitants that live within areas served by a mobile cellular signal, irrespective of whether or not they choose to use it.	Percentage	–		–	Report	Administ-rative Data	Annually	MACRA
Mobile cellular coverage	Percentage of the land area covered by a mobile cellular signal	Percentage	–		–		Administ-rative Data		MACRA
No. of Households with a fixed line telephone		Number	–		–	Report	Administ-rative Data	Annually	MACRA
No. of Households with a mobile telephone		Number	–		–		Administ-rative Data		MACRA
Radio coverage	Percentage of land area with Radio Coverage	Percentage	75		75	Report	Administ-rative Data	Annually	MACRA
TV coverage	Percentage of land area with TV coverage	Percentage	40		40	Report	Administ-rative Data	Annually	MACRA

Contd...

Table 4.1–Contd...

Performance Indicator	*Proposed Definition*	*Measuring Unit*	*Baseline 2010*	*Target 2011*	*Actual 2011*	*Data Sources/ MoV*	*Approach/ Metho-dology of Data Collection*	*Schedule/ Frequency*	*Person/ Entity Responsible*
No. of Telecentres	Count of Telecentres	Number	18		33	Report	Administ-rative Data	Annually	MACRA
No. of Graduating students in ICT at tertiary	Headcount of students having pursued ICT as a course at tertiary level	Number	–		–	Report	Administ-rative Data	Annually	E-Govern-ment
No. of people who are ICT literate	Number of people with ability to use digital technology, commu-nication tools, and/or networks to define access, manage, integrate, evaluate, create, and communicate information ethically and legally in order to function in a knowledge society	Number	–		–	Report	Administ-rative Data	Annually	E-Govern-ment
No. of Schools teaching computer studies at pre-school level, primary, secondary and tertiary	Headcount of schools teaching ICT at various levels	Number	–		–	Report	Administ rative Data	Annually	E-Govern-ment

- ☆ Teledensity has been increasing from 8.15 per cent (2007/2008) to 17.75 per cent (2009/2010) and increased further to 18.32 per cent in 2010/2011. Fixed line density increased from 0.85 per cent in 2010 to 1.19 per cent in 2011, and mobile density also increased from 16.9 per cent in 2010 to 17.13 per cent in 2011. This means there was an increase in access to telephony services per million inhabitants.
- ☆ The number of internet subscribers increased from 2.26 per cent in 2009/2010 to 3.33 per cent in 2010/2011.
- ☆ Both television and radio coverage remained unchanged in 2010/2011.
- ☆ The number of internet subscribers increased from 2.26 per cent in 2009/2010 to 3.33 per cent in 2010/2011, significant increase of 1.07 per cent.
- ☆ It must be noted that there is no data on some indicators as they have not been determined.

It was MACRA's intention that the telecenters should cater to the needs of diverse communities of users by creating their own content according to the requirements. The staff at the telecenters are expected to be knowledgeable so they would be able to customize their home pages and feed the machines with local content.

For the telecenter initiative, MACRA provides rural communities with access to new information and communication technologies (ICT) such as telephones and internet. MACRA equips the people with tools, skills and information needed to compete on an economic level with others in the country and throughout the world. MACRA acts as centers for information distribution and access to different groups in the rural areas. To promote sustainability, MACRA embarked on a public-private partnership to develop and manage the telecentres.

MACRA intended that the management committee oftelecentre will have 10 members, in which 50 per cent are women, and will comprise representatives from: catchment villages, Ministry of Information and Tourism/MACRA, implementing agency and private sector partners.

Telecenter Models in Malawi

Local Management Committee Model

The telecenters are established in the remotest parts of the country, and are run and managed by the communities which choose a committee locally. Scharffenberger (2007) noted the goodness of community telecenters, that there is high level of internet users in the small communities due to two factors:

1. Residents in small communities are more likely to be aware of the existence of telecenters in their locality than those living in larger communities, and thus they would utilize the centers; and
2. The few number of activities available to residents in smaller communities makes telecenters more popular than in large communities with various activities, resulting fewer people utilize the services (Ibid).

Local community Telecenters seek for local development; they often aim to disseminate information, decentralize services, and encourage civic participation, in

Figure 4.1: Lupaso Telecenter: One of the Locally Managed Telecenters in the Northern Tip of Malawi.

addition to providing public ICT access. There is need for these rural telecenters to encourage women to access them.

Entrepreneurship Telecent Model

The telecenters operated by entrepreneurs who meet certain conditions to operate the centres at a subsidized fee. There are currently 8 of these in Malawi.

Postal Telecenter Model

The telecenters are run through the Malawi Postal Corporation (MPC), the organization that runs the country's only postal services. The agreement with MACRA is that MPC should provide space in their Post Offices while MACRA provides computers and internet access. These have been placed in semi-urban centers and they operate on a commercial basis with computers, internet, fax, pay phone, photocopier and scanner. Most of the people who use the service are those who are already computer literate, but some of those who are illiterate still come to access services like printing and photocopying through the help of the staff who work there.

Operator Managed Model

The centers are purely commercial, operated by commercial operators who met certain pre conditions. Currently, there are 10 operators in the country.

School/College Model

MACRA and the World Bank entered into a partnership to enhance the ICT capacity in schools and colleges. The original agreement was that apart from the Colleges and schools the telecenters can also be used by communities. The management of the communities within schools and colleges proved challenging so it was agreed that they will be solely used by the schools.

The general public perception that the centers are within schools negatively impacted their awareness of the services and functions available to them. The perception was created because the telecenters were physically located at an academic facility and thus appeared to be directly associated with students. This made the general public feel intimidated that they would be misfits in that academic establishment. Therefore, these centers have not received much local support.

Target Beneficiaries and Relevant Content

The MACRA's original plan was that a number of sectors will be involved to maximize the benefits as shown in Table 4.2.

Main Constraints of Telecenters

- High cost of internet bandwidth in rural areas
- Unavailability of ICT infrastructure in most rural parts of the country
- Lack of local content (Language and content)
- ICT consumables and ICT technical skills are not within reach
- Insufficient or unavailable professionally qualified human resources to assist ICT integration in the rural areas

Table 4.2

Target Beneficiaries	*Access to*	*Informaton Source and Content Provider*
Small holder farmers	Market Information	Ministry of Agriculture
	Commodity prices	Private Sector
	Agro processing methods	
	Micro finance information	NGOs
Students and youth	Exam results and selection	Ministry of Education
	Distance Education information	Ministry of Health
	Course material	
	Training opportunities	
	Online courses	
	Skills training	
	HIV/AIDS prevention information	
NGOs	Efficient dissemination of information to target beneficiaries	NGOs
	Business information	
	Training materials	
	Community development needs	
	People's rights	
Women	Health information	Ministry of Health
	Economic empowerment information	Ministry of Education
	Linkages with other women organizations	Ministry of Gender
Government Offices	Access to Government forms and documents	Central Government

- ☆ Unregulated ICT training facilities, no standards set in ICT education.
- ☆ Low level of ICT awareness as an enabler for economic development, even at government and administrative levels.

Opportunities

There is Government political will:

- ☆ Tax Removal on internet bandwidth
- ☆ Malawi Government Economy Recovery Plan (ERP) recognizes that the use of ICT could bring economic growth
- ☆ The communication Act is being reviewed
- ☆ Electronic Transaction Bill is being formulated
- ☆ Skilled labour market: Inclusion of ICT in academic syllabus
- ☆ Network of stakeholders: constant consultation on national issues

Findings in the Telecenters

A small survey was conducted with 6 telecenters, to see whether they are doing anything to deliberately involve women in ICT activities. This consequently will improve women's lives by contributing to their socio-economic development. The sample includes each of the models namely: Local Management Committee, Entrepreneurship, Managed by Malawi Postal Corporation, Managed by operator and School/College. One model had two samples because in the whole group these two are the only ones managed solely by a group of women. The Table 4.3 shows the sample.

Table 4.3

Telecenter Model	*Name of Telecenter*
Local Management Committee	Lupaso
Enterpreneurship	Limbika TelecenterNkhotakota
MPC	Dowa Post Office Telecenter
Operator	Ntcheu
School/College	Lakeview SDA College

A small data collection tool that comprise of 8 key questions was developed as below and the findings of the survey are shown in Table 4.4.

1. Are there any women managing the telecenter? (percentage)
2. How many women access the telecenters in a day? (percentage of patronage)
3. Are there any organizations working with special women groups in the telecenters?
4. Has local content been produced to suit the needs of the community?
5. Does the telecenter have qualified staffs who offer computer lessons to women and other groups?

6. Has awareness been created to make the public know the existence of the telecenter?
7. What are the most common services accessed at the telecenter?
8. Is power supply available at all times?

Table 4.4: Findings to the Survey Questions

Questerion	*NKHOTA-KOTA*	*LUPASO*	*LIMBIKA*	*LAKEVIEW*	*NTCHEU*	*DOWA*
1	100 per cent	50 per cent	100 per cent	30 per cent	10 per cent	50 per cent
2	16 per cent	42 per cent	28 per cent	48 per cent	36 per cent	19 per cent
3	Yes	No	No	No	No	No
4	None	Scanty	None	None	None	None
5	Yes	Yes	Yes	Yes	Yes	Yes
6	Yes- through Community radio	Yes posters	Yes Posters	School bulletin	Yes posters	Yes posters
7	Internet	Printing	Photocopy	Internet	Internet	Photocopy
8	Good	Good	Unreliable	Fair	Fair	Intermittent

General Findings

The general findings of the survey indicate:

Women Management

There is an effort to have women managing the telecenters as such it is a good attempt for increased accessibility to women. In most cases, there are women managers, because the pre condition set by MACRA and the other funders of the telecenters stated that 50 per cent of the managers should be women. It is therefore evident that somehow women are involved in ICT in the rural areas.

Women Access

The access and utilization of ICT services among rural women is still minimal. Regardless the fact that some of the centers are managed by women only, however, the number of women accessing the centers is discouraging. Women continue to be the least to access the ICT regardless of who owns the centers. This is evident and is due to non availability of special programmes to attract women.

Special Women Activities

There are no special women's activities taking place at the telecenters. Nkhotakota is the only exception, since they organize computer training for female Primary School teachers alongside other groups like school children and library workers. Other than that, no efforts by Local NGOs or the Ministry of Gender to mobilize women, so that their capacity will be enhanced to allow them to access to relevant information pertaining to their situation in their community.

Local Content

There has not been muchdone in the telecenters to localize language and the content. Nkhotakota and Lupaso have made some effort by customizing their home pages and using their local languages on the screen saver and the home page. Other than that, the centers have not been fed with deliberate local content that can impact on the local community, especially on local women.

Qualified Staff

It was found that all the sampled telecenters have qualified staff who can teach basic ICT to the local people, although that is not the general practice. Except for Nkhotakota and Lupaso where the staff has made efforts to teach people, little has been done to transfer the computer knowledge to the local masses to make them computer literate so that they can access relevant information either by themselves or in groups.

Awareness

In all the telecenters sampled, there is awareness of the existence of the telecenter. In the cases of Nkhotakota, where there is a community radio close by, the news was spread adequately and the response from the general public was quiet encouraging. In most cases posters were used, but the biggest awareness was created during the launching ceremonies of the centers, which mostly involved politicians or administrators of the local districts. Although opening of telecenters by politicians gave a wrong perception to public, but it created the much needed awareness.

Most Common Service

The most common service sought is the internet. Most people use it just for social interaction but it was discovered in college that apart from social networks, students use internet for research and to access course materials. Photocopy service comes next followed by printing service. The secretarial services are sought for CVs and job application letters.

Power Supply

In most of the telecenters, the electricity intermittent supply has worked negatively. This indicated that, at most, electricity problems are experienced after working hours when the centers have closed. Services are therefore interrupted due to the numerous black outs that take place.

Observations

At the outset it is very clear that regardless of the fact that the Government has made efforts to increase universal access of ICTs in the rural areas, the implementation of the programme still leaves a lot to be desired. The usual challenges of ICT access and utilization come into play.

Challenges

- First and foremost, there are very few telecenters that have been established in the very remote areas and the most obvious reason is that it is very difficult to bring electricity in places which are not connected by the local provider ESCOM and as such, the GoM takes advantage of the areas that already have electricity to avoid extra expenses.

- ✰ **Almost nonexistent computer usage in the rural areas:** Apart from the few rural areas which have got Telecenters, there is almost zero existence of computers and other ICT gadgets in the rural areas, let alone people who are able to use them. The only form of ICT available are mobile phones.
- ✰ **Lack of financial resources:** For the average Malawian the amount of money that people are expected to pay in the Telecenters is too high even at the subsidized rate. People would rather spend their money on food or household products such as soap. Even to spend time at the telecenter, they think is a waste of time as they would rather be in their fields working or at the market trading and get an income at the end of the day.
- ✰ **Staff attitude and qualification**: Most of the staff at the telecenters do not want to be in the position to serve; regardless of the fact that these are also social ventures, most staff are in it for the money. On top of that, there are only a few of the staff working in these telecenters that are adequately qualified to get to the level of transferring knowledge to someone who doesn't know how to work with the ICT gadgets.
- ✰ **Culture:** The Malawian culture insists that the woman's place is in the home, especially in the rural areas. Therefore, a woman who will be seen to go out of the home to mingle with men and young people will not be understood.
- ✰ **Location:** Most of the telecenters in Malawi are located in sophisticated institutions where the local woman is not accustomed to. It therefore becomes intimidating for the local woman to find herself in a locality that is more advanced than her own.
- ✰ There is **limited local access and international connectivity** required to provide cost effective broadband to the rural areas.
- ✰ **Illiteracy**: Most Malawian rural women are illiterate and their illiteracy therefore makes it difficult for them to be able to learn how to use the ICT gadgets.
- ✰ **Politics mixed with the running of the Telecenters**: There are a lot of issues involved in running of these telecenters which makes the managers concentrate on other things other than thinking of making provisions for the rural woman.

Recommendations

This paper makes the following recommendations to improve the access and use of ICT among rural Malawian women so that they improve socio-economically and improve their lives and that of their communities (social economic development).

1. The Government of Malawi should consider taking telecenters to the remotest areas because once the rural women discovers that there is such a

facility in their locality they will understand it is for them and may feel obliged to try it.

2. Before telecenters are established, there is need to create awareness and understanding amongst the masses so that they know that this is a technology that once utilized properly can change their lives and empower them economically and socially. Deliberate efforts should be made to demystify ICT and make the people understand that it will bring development.
3. Involvement of local leaders is crucial. It is necessary to involve the grassroots administration first, because once they have been made to understand a concept, It has been easy for them to convince the locals to follow suit. There are numerous examples in Malawi of how communities have accepted programmes because the Government got a buy in from the local leaders first.
4. GoM should make deliberate efforts to include women education in ICT by including it in the relevant instruments.
5. Culture is a barrier to women accessing and using ICT; again it is necessary to involve local leaders to free and accept women to go to places where they can access such services.
6. Non Governmental organizations especially those dealing in women's rights promotion and those dealing in the advancement of women, for instance the Society for Advancement of Women (SAW) and Coalition of Women living with HIV and AIDS (COWLHA) should get involved to mobilize women as their lives will be improved if they access pertinent information through ICT.
7. The GoM through, the Ministry of Education, the Malawi Institute of Education (MIE) and the University of Malawi should come up with localized computer software in local languages for use in the rural areas.
8. GoM and the funders of the ICT projects should find a way of having special groups of women access ICT free of charge as poverty is a very big factor, as far as the access of ICT by the rural women is concerned.
9. Storytelling (Reilly and Gomez, 2001, p. 7; Gomez and Hunt, 1999, p. 2) should be employed as one of the ways of attracting the women to access ICTs.
10. Managers of the telecenters should use creative ways of using information that has been accessed using ICT for instance, disseminating collective knowledge acquired through email and/or the Internet and distributing through bulletin boards, links to community radio, village meetings etc.

Conclusions

The Affordable use of ICTs in Africa is critical, not for the mere ability to conduct social intercourse by electronic means, or to improve the delivery of governmental and business services to isolated communities; rather, it is central to the core objective of empowering people through literacy, education, knowledge, employable skills, poverty reduction and wealth creation (CTO, Commonwealth African Rural Connectivity Initiative, 2008).

Acronyms

COWLHA: Coalition of Women Living with HIV and AIDS
ERP: Economic Recovery Programme
ESCOM: Electricity Supply Commission of Malawi
FISP: Farm Input Subsidy Programme
GWAN: Government Wide Area Network
GoM: Government of Malawi
ICT: Information and Communication Technology
ICTSRDP: ICT for Sustainable Rural Development Project
LMC: Local Management Committee
MACRA: Malawi Communication Regulatory Authority
MGDS: Malawi Growth and Development Strategy
MICE: Ministry of Information and Civic Education
MPC: Malawi Postal Corporation
MPCT: Multipurpose Community Telecentre
NGO: Non Governmental Organization
PC: Privatization Commission
PIAP: Public ICT Access Points
PPP: Public Private Partnership
RCIPMW: Regional Communication Infrastructure Project Malawi
SAW: Society for the Advancement of Women

References

1. Dahms, Monica. 1999. For the Educated People Only. Reflections on a Visit to Two Multipurpose Community Telecentres in Uganda, in Telecentre Evaluation: A Global Perspective. Paper published on the World Wide Web and available at

   http://www.idrc.ca/telecentre/evaluation/html/main.html.

2. Gomez, Ricardo, P. Hunt, and E. Lamoureux. 1999. Telecentre Evaluation and Research: A Global Perspective. Paper published on the World Wide Web and available at

   http://www.idrc.ca/telecentre/evaluation/html/main.html

3. Sagna, Olivier. 2000. Information and Communication Technologies and Social Development in Senegal: An Overview. Available at http://www.unrisd.org/cgibin/dnld1.pl?filename=infotech/sagnaeng.pdf:385.5k and thispage

=infotech/publicat/publ.htm and filetitle=Information+and+ Communications+Technologies+and+Social+ Development+in+Senegal

4. Scharffenberger, George. 1999. Telecentre Evaluation Methods and Instruments: What Works and why?, in Telecentre Evaluation: A Global Perspective. Paper published on the world wide web and available at http://www.idrc.ca/telecentre/evaluation/html/main.html
5. Government of Malawi (GoM) Official website www.malawi.gov.mw
6. Malawi Growth and Development Strategy 2011-2016 (MGDS II)
7. Malawi Communication Regulatory Authority (MACRA) website www.macra.org.mw

Chapter 5

Women Empowerment through Information and Communication Technology in Nepal

Parbati Pudasaini

Nepal Academy of Science and Technology,
PO Box 3323, Kathmandu, Nepal
E-mail: prpudasaini@yahoo.com

ABSTRACT

Gender equality has become one of the major burning issues prevalent in the developing countries like Nepal. Equal opportunity for men and women in economic activities is crucial for women empowerment in terms of development of decisive power in different aspects. Women's economic empowerment means greater access to financial resources inside and outside the household and reducing vulnerability of poor women to crisis. As the power to retain income and use it at the decision it provides equal resources at the household level. The most important issue is to identify the exact situation and real problems related to getting access to education and opportunity for employment. It is obvious that it is possible to uplift society only when the mix of human resources is used as human capital. In Nepal more than 50 per cent of the total population is women and around 46 per cent of the economically active population is women. Traditionally, women in Nepal have been contributing more actively than their male counterparts, mainly in domestic and household works and agriculture. Still, the economic contributions made by women at household level or in agricultural sector are not recognized by the society and as a result of which most Nepalese women suffer from unequal power sharing with men. Information and Communication Technology (ICT) has emerged as an effective vehicle for social, political and economic change through the widespread access to information to even the marginalized cross-section of our society including women as a result the issue of women empowerment has been brought to the forefront of national agenda. Nepal is now at the juncture of its history and is going to draft a new constitution that is expected to ensure the social inclusion, the major agenda of which will be women empowerment in all sectors and at all levels.

Keywords: *Women empowerment, Nepal, Equal participation, Educational opportunity, Domestic Violence, Women Trafficking, Community Media, Employment.*

Introduction

There is no argument against the empowerment and autonomy of women in Nepal and it has been recognized that the social, economic and political empowerment of women is essential for their sustainable development in all areas of life. Thus it has been well accepted that women's empowerment and full participation on the basis of equality in all spheres of society including in decision making process and access to power are fundamental to the achievement of equality development and peace. Accordingly, appropriate policies and programmes have been formulated to address women's development issues and problems in various ways at national as well as local levels. However, due to the absence of effective mechanism and political commitment to implementing them women's concerns and needs tend to be marginalized and lost during the course of implementation.

The biggest problem on empowering women in Nepal is the wide gap between policies, plans and programmes on the one hand and their actual implementation on the other. This is because of weak governance, lack of capacity for gender analysis and planning, insufficient efforts to include women's representation in decision making process and lack of interest and commitment to implementing women's development programmes in local levels. Also, the lack of understanding on the part of decision makers, less attention towards women's empowerment and emphasis on women's physical rather than qualitative participation are the other crucial factors for the weak implementation of women's development plans and programmes.

Human Trafficking, especially of women and children, is a serious social crime in Nepal. No exact data on human trafficking is found in any country and the same is true for Nepal. Human trafficking in Nepal began with women and children for sexual exploitation. Increasing unemployment in the nation and growing attraction for foreign employment has become a subject of manipulation for human traffickers in Nepal.

Women Empowerment

Policy Framework

The Government of Nepal adopts special policies and guidelines to encourage women's participation in all spheres of employment. Government of Nepal is dedicated to create more training for skills development, which paves the way for women towards opportunities for income generation and self-employment in both urban and rural areas. Studies to simplify institutional credit facilities and collateral-free loans are offered to poor and uneducated women, particularly in rural areas.

Awareness programmes are conducted to reduce prejudicial acts against women and increase family support to women's employment and also to uplift women's social status. Mass awareness campaigns are launched against domestic violence and discriminatory social norms and values. But, women are virtually absent in the important news coverage related to Government's policies and programmes, whether it is transmitted through press or aired on radio or television. Therefore, women lack access to information they need and to which they have right. Only information can help them answer questions affecting their daily lives, problems and needs. Women empowerment is not possible without giving them adequate access to information related to their rights.

Roles of Media

History of Women in Journalism

The history of journalism started in Nepal in 1901 with the publication of the **Gorkhapatra,** the first national daily. In this history of more than one hundred years of Journalism many political ups and downs have taken place but still the country could not develop the journalism accordingly. When the autocratic Rana rule of 104 years was overthrown in 1950 the information and communication sector tried to take steps ahead little bit. In that time the journalism was only for the welfare of the ruler, the then head of Government as a result of which professionalism in journalism could not be developed at that time.

Involvement of women in journalism does not have that long history. In 1951, when Sadhana Pradhan and Kamakchha Devi Published the first women monthly paper, **Mahila**, then, the history of women journalism began in Nepal. In that **"Mahila"** (Women) monthly paper Sadhana Pradhan and Kamakchha Devi were editors and Seelbanti, Madhukar and Pramila were co-editors and Mayadevi was manager.

Priyamba Sharma and Laxmi Devi also put in their efforts to continue what was initiated by the pioneer women journalists. In October 1951 Priyamba Sharma published a monthly paper **"Prava**" in which Priyamba Sharma was Editor and Laxmi Devi Sub-editor. Similarly in 1952 under the editorship of Kunti Devi a monthly paper entitled "**Prativa"** started to be published. Likewise, in 1953 Rama Devi came up with "**Janabikash"** monthly in which she was editor as well as publisher. In fact, Rama Devi can be called the first Nepali lady who came up with the pure Journalistic view and interest because all the women in journalism before her were involved as an editor or publisher from some women organization or from some schools' manifesto group or from some political party organizations. The "Janabikash" was published up to 12 issues whereas earlier ones got stopped after the publication of their first or second issue.

In the later years Rita, Kalpana, Chanda, Bijaya Laxmi Malla, Laxmi Devi Chitrakar, Sashikala Sharma, Sarojani Manandhar and Laxmi Ranjitkar emerged as women journalists. In 1954 the monthly paper "**Bhor**" was published in the editorship of Rita and Kalpana. The same year Chanda published "**Rangamanch**" in her editorship. Likewise, in 1958, Bijaya Laxmi Malla and Laxmi Devi Chitrakar published "**Ganbikash**" monthly in their editorship. In the same year **"Suwasnimanchhe**" (the women) was also published in the editorship of Shashikala Sharma. Only this **"Sawasnimanchhe"** could continue for long time. In the beginning, it was quarterly and later on it started to be published bi-monthly. After the success story of the "**Suwasnimanchhe"** Sarojani Manandhar and Laxmi Ranjitkar published a fortnightly paper called "**Chetana"**. In this way Shashikala Sharma played a pivotal role to build confidence among the women that they too can produce a very good and important newspaper/magazine in difficulties also.

In the list of pioneer Nepali Women Journalists, the name of Umadevi is also very important. In 1962 "**Mahila Bolchhin"** weekly was published in the editorship of Umadevi. This was first weekly with woman as an editor and was regularly published until 1990. The leading women journalists of Nepal are Madalasha Shing, Santa Shrestha, Rasmirajya Laxmi Rana, Prava Basnet, Ranjana Sharma, Mithu Devi, Gauri K.C., Krishana Tamrakar, Bina Gurung, Ritaraj Gurung, Rani Gurung, Nupur Bhattacharya, Dhana Lama, Surya Kumari Pant, Kalyani Rimal, Sarita Bhatta, Bharati Silwal of Radio Nepal, who can never be ignored while taking an account of the development of ICT in Nepal.

Community Multimedia Centers

In the country like Nepal with poor resources information is often tightly controlled and restricted. But, community multimedia centers have dramatically increased the flow of information into thousands of households. From the global reach of internet with local media in local languages and dialects, Community Multimedia Centers (CMCs) make it possible for women to access new types of information on health, agriculture and all the aspects of life. CMCs enable people, especially women, to communicate their ideas and problems to participate in public discussions and in the lives of the peoples of their community.

Modern Media Based on ICT

From 1985, when the Nepal Television started its regular transmission, the number of women journalists increased enthusiastically. After the political change

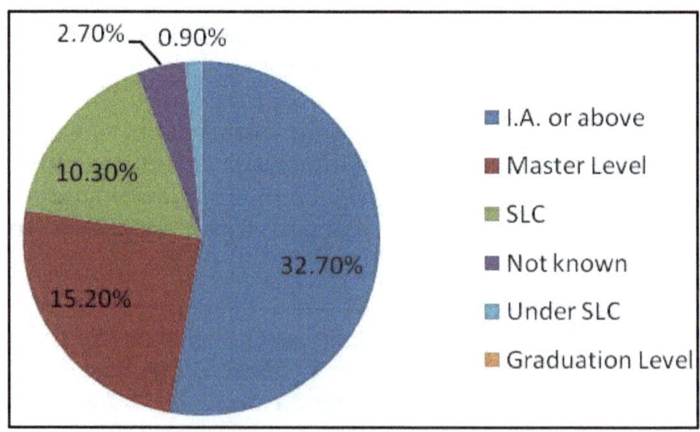

Figure 5.1: Pie Chart Showing the Composition of Academic Qualifications of the Women in Media.

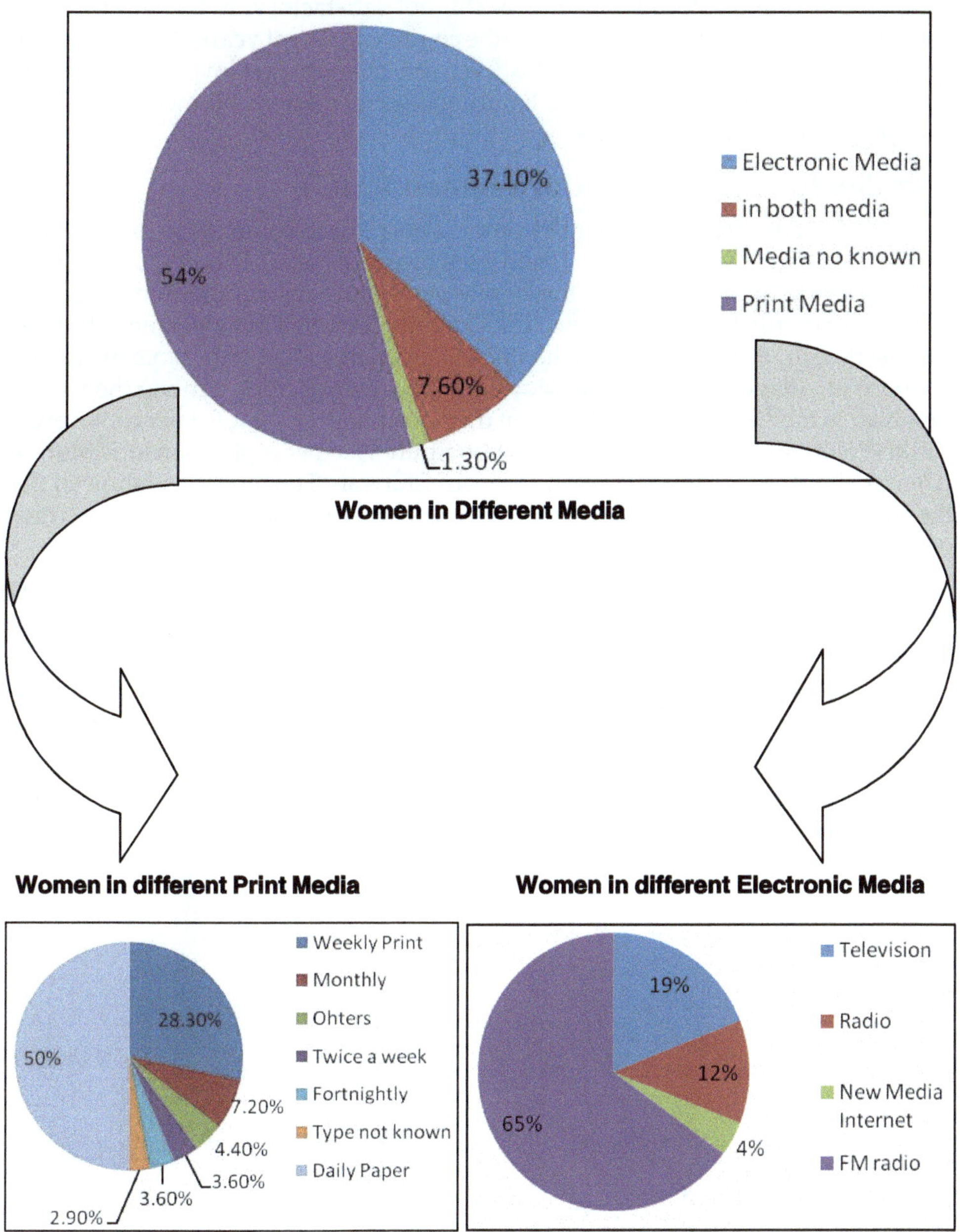

Figure 5.2: Pie Charts Showing Involvement of Women in different Types of Media Operating in Nepal.

in 1992 the professional journalism developed in Nepal. Likewise, women journalists also came ahead little bit more. Although the number of women involved in journal compared to similar number of men is still not satisfactory. Even with the rapid growth of electronic and print media in the country in recent years, the involvement of women in this field is still very little. Even the ones who get involved in this field cannot stay for long because of their family and social matters. Women get attracted towards this field quickly and they quit the profession too very quickly.

Hurdles for Invovlement of Women in ICT

It is obvious that the number of women is very less in comparison to that of men and that too short-lived in electronic and print media of Nepal. In order to identify the real problem behind this fact, the forum of women journalists and communicators of Nepal, commonly known as "Sancharika Samuha" carried out a detailed study for the first time in 2001. The report of the study showed that most of the women in media are born in villages and later they started their career in cities. Among the women involved in media, 25.9 per cent were of the Kathmandu valley, 9.8 per cent were of Kaski (Pokhara), 9.4 per cent of Banke (Nepalgunj), 7.6 per cent of Sunsuri/Morang (Dharan, Biratnagar) and 7.1 per cent were of Chitwan. The report also showed that the educational levels of women in media were found to be satisfactory of which most of them had completed undergraduate school or even more.

The compositions of the women involved in different types of media are shown in the pie charts as shown in Figure 5.2 and the comparison of the different positions occupied by women in different media houses of Nepal is shown in Figure 5.3.

The same forum called "Sancharika Samuha" did carry out another research in Kathmandu Valley in 2011, which covered 23 presses, 9 Television Channels and 14

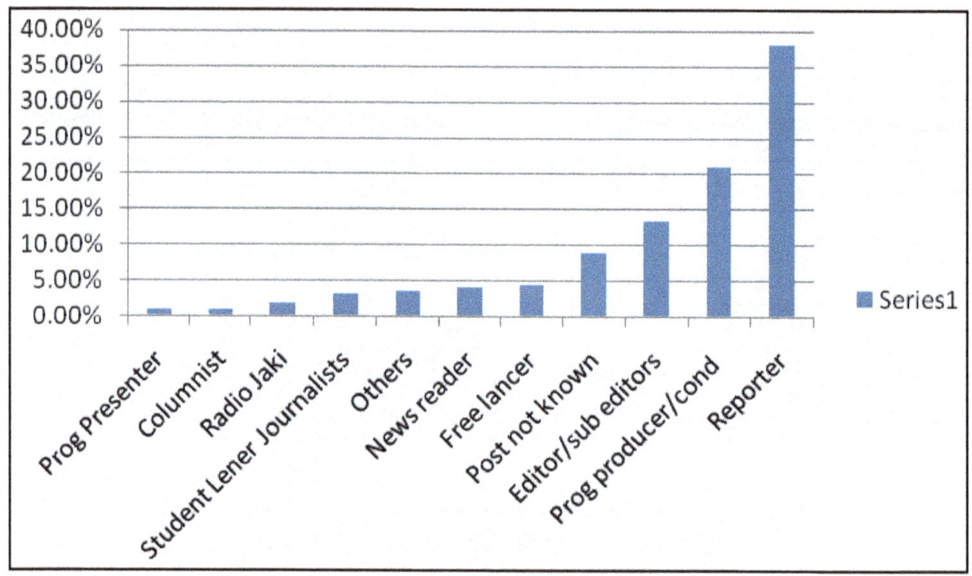

Figure 5.3: Bar Diagram Showing the Comparison of the Percentage of Women Working in different Positions within Media Houses.

radio channels. In those 46 media houses a total of 1,815 persons were involved out of which only 446 were women. In the country where more than 50 per cent population is of women, the poor representation of women (only 24 per cent) in media shows plenty of room for improvement. During this study a total of 152 women were directly interviewed out of which 43 per cent were found to be working in Radio, 32 per cent were in print media and 26 per cent were in Television as shown in Figure 5.4.

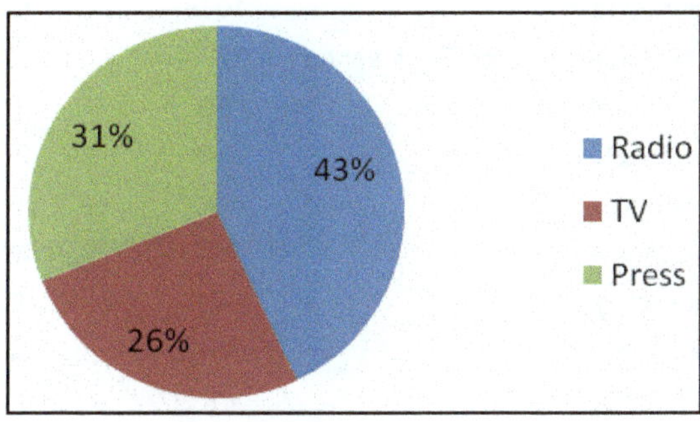

Figure 5.4: Pie Chart Showing the Comparison of Women Involved in different Media Sectors.

Results indicated that Radio is the first preference of women in electronic media because it is easier and reachable for them whereas the print media with the history of more than 100 years is in their second choice. Though Television has a very short history of only 27 years in Nepal it falls in the third choice of the women.

The Table 5.1 presents a comparative data of the men and women engineers of different disciplines registered in Nepal Engineering Council until 2012.

Table 5.1: Comparison of Number of Male and Female Engineers Registered in Nepal Engineering Council until 2012

Sl.No.	*Engineering Faculty*	*Male*	*Female*	*Total*
1.	Aeronautical Engineering	123	2	125
2.	Agriculture Engineering	160	21	181
3.	Agri-irrigation Engineering	22	0	22
4.	Architect	445	521	966
5.	Automation Engineering	4	0	4
6.	Automobile Engineering	6	0	6
7.	Avionics Engineering	0	0	0
8.	Bachelor of Urban and Physical Planning	5	0	5
9.	Bio-medical Engineering	67	29	96
10.	Chemical Engineering	46	4	50
11.	Civil Aviation Operation and Engineering	0	0	0
12.	Civil Engineering	7234	540	7774

Contd...

Table 5.1–Contd...

Sl.No.	*Engineering Faculty*	*Male*	*Female*	*Total*
13.	Computer Engineering	2011	526	2537
14.	Earthquake Engineering	0	0	0
15.	Electrical and Electronics Engineering	674	78	752
16.	Electrical Engineering	1048	56	1104
17.	Electronics and Communication Engineering	2739	422	3161
18.	Electronics and Telecommunication	63	3	66
19.	Electronics Engineering	239	21	260
20.	Energy Engineering	0	0	0
21.	Environmental Engineering	68	20	88
22.	Forestry Engineering	7	0	7
23.	Geology Engineering	20	0	20
24.	Geotechnical Engineering	1	0	1
25.	Industrial Electronics	1	0	1
26.	Industrial Engineering	125	4	129
27.	Information Technology	199	36	235
28.	Information Technology and Telecommunication Engineering	8	3	11
29.	Instrumentation Engineering	0	1	1
30.	Manufacturing Science and Engineering	0	0	0
31.	Mechanical Engineering	1237	28	1265
32.	Metallurgical Engineering	6	0	6
33.	Meteorology Engineering	0	0	0
34.	Metrology Engineering	0	0	0
35.	Mining Engineering	38	0	38
36.	Radio Engineering	16	0	16
37.	Software Engineering	46	141	60
38.	Sound and Video Engineering	0	0	0
39.	Survey Engineering	11	0	11
40.	System Engineering	25	5	30
41.	Textile Engineering	13	2	15
42.	Biotechnology	8	1	9
43.	Non Nepali Registered Engineers	18	0	18
Total	16732	2234	19066	

Being optimistic towards the future, in the pretext of revolution in ICT globally and new inclusive constitution being drafted locally, Nepalese women will definitely come forward to be competitive enough with men every sector including ICT in near future.

Conclusions

In the developing countries like Nepal information is often tightly controlled, in contrary to which the Community multimedia centers (CMCs) of Nepal have dramatically increased the flow of information. By combing the global reach of the internet with local media in local languages and dialects, CMCS make it possible for people to access new types of information on health agriculture and on a range of other topics.

Making access to education to all the girls of remote areas of Nepal is the major challenge in Nepal. A central effort through CMCs is to innovate ways to support and supplement the education of poor girls along with boys also. Using both radio and internet, CMCs have great potential to deliver educational curriculum especially in remote areas.

Many CMCs in different parts of Nepal work specifically with youth from poor families, minority groups and traditionally marginalized caste groups. Youth are always interested in learning the use and value of computers. Many young people show wonderful ability to learn the functions of new technologies not being limited only to computers and internet but also the sophisticated software and networking. The CMCs encourage the girls and boys to be a practical human link between their families.

Unemployment and underemployment are very high in Nepal especially in youth. CMCs not only provide the young people with new skills and access to resources but also offer a place in which they can put their skills to work.

In the context of Nepal, women constitute more than 50 per cent of the total population. However, their representation in every sector of development is very limited. Therefore, there is a need to understand the role of women in the economy of today. Some efforts have been made to empower women and their participation in every sector but that are not enough. There are many legal provision and policies to improve the participation of women, but in reality, social barriers and other forms of resistance do not allow them to participate in the required manner. After the promulgation of long awaited inclusive new constitution of Nepal in near future equal participation of men and women in all sectors and at all levels can be expected to come as a reality.

Acknowledgement

1. Sancharika Samuha – Nepal
2. Ministry of Women Children and Social Welfare, Government of Nepal
3. NAM S&T Centre – New Delhi
4. Prof. Dr. Nirupama Prakash

References

1. Survey report of Sancharika Samuha.
2. Annual report of Ministry of Women Children and Social Welfare Government of Nepal.
3. Different Website of women and gender related organizations of Nepal.

Chapter 6

Social Security of Women in Developing Countries through ICT

Isha Sehgal[1], Pinky Singh[2] and Parul Sehgal[3]

Research Assistants,
Centre for Science and Technology of the Non-Aligned and Other Developing Countries (NAM S&T Centre), New Delhi, India
E-mail: [1]sehgalisha305@gmail.com; [2]pinky.bharta@gmail.com; [3]parulsehgal1989@gmail.com

ABSTRACT

In spite of considerable progress in many developing countries, women are still considered to be a marginalised section of the society and face social discrimination and multifarious problems like lack of access to education, health and economic independence, inadequate safety and security, violence, etc. Information and Communication Technology (ICT) has emerged as an effective tool for bringing about a revolution in the societies world over and offers an opportunity to people to rise against all odds. Considering the situation of women in economic, social, education and health sectors where they are highly discriminated and face gender inequality, ICT can prove to be an empowering factor for improving their status while offering equal opportunities as men. Through proper communication facilities and technology, women can get access to information leading to awareness while enhancing their knowledge and skills in a particular domain. ICT holds a very important place when it comes to ensure social security of women in the developing countries. Violence against women is one of the major issues which is prevalent in our society for an exceptionally long period of time and needs to be addressed with a modern approach to put an end to it. Application of ICT can bring about a positive change in this regard with the help of several technological innovations like internet, mobile phones, social media, radio and television, etc. which help in women empowerment and providing adequate social security to women in all spheres of life through capacity building, affordable health services, employment opportunities, and improved social, economic and gender equality.

Keywords: *Information, Communication, Gender equality, Opportunities, Social security, Technology, Violence.*

Introduction

Information and Communication Technology (ICT) is an extremely powerful tool and plays a key role in improving the lives of people in many ways while uplifting their standards, reducing poverty, enhancing productivity thereby leading to socio-economic growth and development of a country. This implies that ICT literacy has to be an integral part of everyone's life so as to remain updated with latest technologies, innovations and enhance their future with better opportunities. ICT forms a platform for information dissemination and it must be utilized properly as it can benefit only those who have an access to it. So, if there is no access due to economical, social or physical conditions then it will only pose a drawback for the people and hence for the nation because it will cause them to fall back and remain incompetent technologically. Everyone has a right to information and ICT which is a powerful medium, allows for sharing of information that is fast, accurate, location independent and economically feasible for people all over the world especially in the developing nations that need to advance their information technology and hold a very high percentage of people who are deprived of this wonderful and extremely useful tool. Therefore, the government needs to place a focus on the acquisition and use of ICT and devise various strategies for incorporating it at a larger scale while covering people from all sections of the society irrespective of rich or poor without any discrimination based on caste, sex and creed. There are different modes of ICTs which include new advanced technologies like Internet, mobile phones, tablets, social media, radio, television, etc. that have proven their potential to bring about a revolution and reform the existing system of various sectors be it education, research, media, healthcare, travel, government, etc. In this current century, where we increasingly rely on ICT, there is an urgent need of modern infrastructure and highly skilled ICT workforce required in large numbers so as to meet the increasing demand of modern economies.

According to Crede and Mansell (1998), ICTs are crucially important for sustainable development in developing countries. Thioune (2003) also noted that for the past two decades most developed countries have witnessed significant changes that can be traced to ICTs. These multi-dimensional changes have been observed in almost all aspects of life: economics, education, communication, and travel. In a technology-driven society, getting information quickly is important for both sender and the receiver. ICTs have made it possible to quickly find and distribute information. Annan (2002) noted that the information society is a way for human capacity to be expanded, built up, nourished, and liberated by giving people access to tools and technologies, with the education and training to use them effectively. There is a unique opportunity to connect and assist those living in the poorest and most isolated regions of the world. Informatization of society is a major hurdle that most nations, especially developing countries, are encountering. The information society or information age is a phenomenon that began after the 1950s, which brings challenges as we seek to integrate and expand the universe of print and multimedia sources. The two terms are often used to describe a cybernetic society in which there is a great dependence on the use of computers and data transmission linkages to generate and transmit information (Bruce, 1995).

Women and ICT

ICT has undoubtedly changed the society and rules our lives but even then there is a significant gap in the way the genders interact with it. Let's throw some light on it and take it to be more precisely, women as compared to men participate less in ICT related activities. They do not utilize it fully and as frequently as men because of various prevalent social and cultural norms while hindering their progress and not allowing the utilization of advanced technology to stay ahead and remain competitive to face various challenges of the global community. This disparity in the form of unequal distribution of ICTs between not only the sexes but also different communities, regions, ethnicities has taken a bad shape and needs to be addressed at the earliest so that it can be handled in the best possible ways.

International Telecommunications Union: ICT Facts and Figures (For the year 2013)

According to the ITU facts and figures, over 2.7 billion people have been using internet so far, which is around 39 per cent of the world's population. In the developing countries, 31 per cent of the population uses internet as compared to 77 per cent population in the developed countries (Figure 6.1).

Acknowledging the gender gap here, it is estimated that men use internet more than women world over and around 41 per cent men use internet as compared to

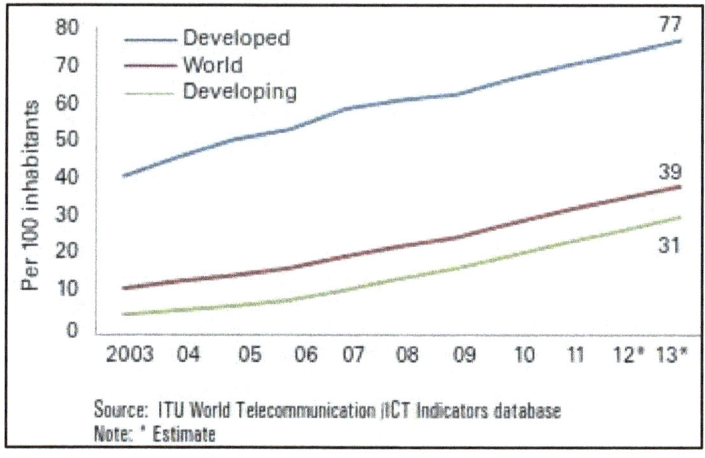

Figure 6.1: ITU Facts and Figures Graph (a).

women with 37 per cent. As per ITU, the developing countries have 980 million males using internet as compared to 826 million females whereas the ratio is comparatively less in the developed countries with 483:475 male to female internet users' ratio (Figure 6.2).

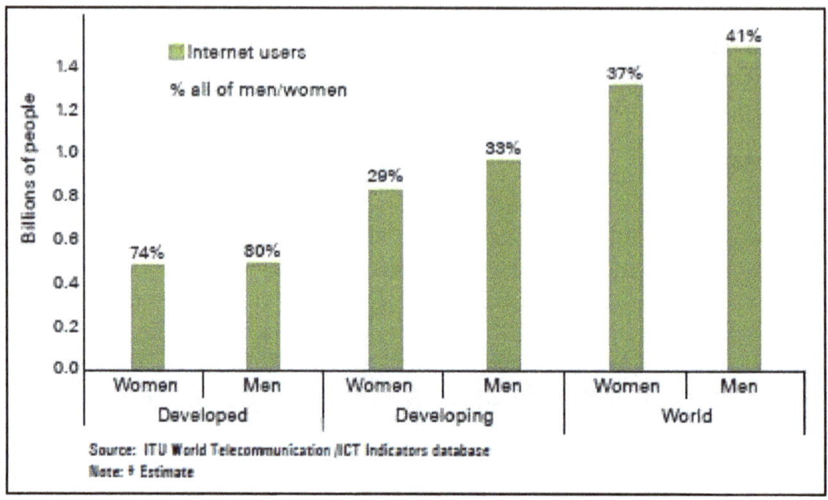

Figure 6.2: ITU Facts and Figures Graph (b).

Social Security of Women and ICT

Social Security means to ensure protection in all arena's of life like providing income at the time of unemployment and in case of retirement or disability, basic healthcare facilities and social services that includes protection against violence, encouragement for their participation in various social and cultural activities, inspiring them to be a decision-maker while also ensuring gender equality. Social security is extremely important especially for women as throughout lifetime they are paid less and their life expectancy is generally higher as compared to men. As reported by the ISSA (International Social Security Association) on International Women's Day celebrated on March 8, 2013, "Research demonstrates that women accumulate social and economic disadvantages over their lives and careers, owing to their occupation of unpaid, low-paid or informal economy work. Women therefore continue to be less often entitled to contributory pension benefits, and their pensions are often significantly lower than those of men due to lower earnings and shorter contribution periods."

Although, the developing countries are progressing as far as their economies are concerned but the situation of women in these countries is distressing and they face lot of obstacles which hinder their path to lead a safe and secure life. This situation has a direct impact on their access to social security protection which puts them at high risk. We may put it as "Social Security is a right for all but enjoyed by only a limited number of people worldwide".

There are a number of problems faced by women in the developing countries and some of the major problems are listed below:

1. Female Foeticide and Infanticide
2. Lack of access to education
3. Less participation in Science and Technology
4. Fewer Employment and Advancement Opportunities
5. Discrimination Against Men- Do not enjoy the same rights as men
6. Women Entrepreneurs face greater hurdles due to various social and cultural barriers in marketing, management and technology skills
7. Rural women face drudgery, gender inequality, lack of awareness, etc. causing them to stay backward socially and economically
8. Less access to better healthcare facilities
9. Women do not feel safe and secure

All the afore-mentioned problems can be tackled through the application of ICT so as to improve the quality of life of women and ensure adequate security to them with the help of several technological innovations like Internet, mobile phones, social media, radio, television etc.

Let us take the role of ICT in solving such important issues.

Girl's Education

The Problem

Millions of girls in the developing countries remain un- or under-privileged as they get none or little formal education. At a very early stage of their lives, they are discriminated on the basis of gender which hampers the development and utilization of their full potential and curb the chances for higher education (Figure 6.3).

What are the causes?

☆ **Lack of family support and Gender Gaps in Education:** This indicates that the mentality of parents is such that they are not at all willing to send

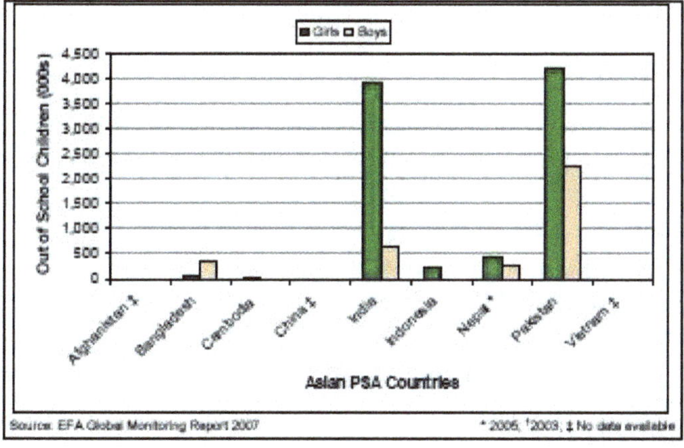

Figure 6.3: Number of Primary-Aged Children Out of School in 2004.

their girl child to school considering that after marriage the girl will leave the house and will be taking care of her husband and children, hence, why should they waste money on her. Rather the emphasis is on the boy child who would stay with the parents and bring home a considerable amount of salary. Therefore, the girls are forced, pushed and persuaded to follow the traditional pattern of the society wherein a girl or a women has the role of taking care of the household activities and looking after her children and other family members.

- ☆ **Accessibility:** This includes the distance of the school from home, lack of schools in the vicinity and unavailability of proper infrastructure. Parents are always worried about sending the girl child alone to any far off areas as there is a sense of insecurity which poses as a restriction for the girls to have equal access to education.
- ☆ **Language Barrier:** As a matter of fact, most of the information available globally is in the English Language which is not understood by everyone especially people staying in the remote areas and poor communities who can generally speak in their local and regional languages, that makes the situation complex.
- ☆ **Various Socio-cultural Practices:** Orthodox attitude of the society towards the female gender is one of the main causes behind the restricted development of the girls or women world over. The societal unawareness of the potential benefits of girl's education adds on to the awful condition. Further, the girls may face sexual assault, violence from different members of the society working at various levels.
- ☆ **Financial Strain:** High levels of poverty, unemployment, food unavailability, sickness along with high cost of education paralysis their efforts to achieve educational qualifications.
- ☆ **Lack of role models:** There is a lack of female role models and female teachers in schools who can set an example for the girls to go for education so as to walk on a path of growth and development without any fear, breaking away all chains of illiteracy and coming out successful in life.

ICT as a Solution

The medium of Information and Communication Technology is extremely helpful in providing a solution to the commonly faced problems by allowing access to a pool of knowledge and information. ICT can provide the following:

- ☆ Use of radio and television for formal and informal training for *e.g.* India with its radio service like All India Radio having more than 200 radio stations and 300 transmitters and television channels like Doordarshan with 600 transmitters, has a tremendous source of information which can help in disseminating high level of knowledge and information while reaching people at regional and national level. (http://wcd.nic.in/research/ict-reporttn.pdf)

- ☆ With the help of basic computer literacy programmes, wherein it will be taught to make use of computer and internet, it would be possible to impart training to teachers through distance learning and deliver jobs to female teachers as well, girls restricted to homes can also make use of this opportunity and learn through distance education and it will also influence the enrolment number of girls in schools. For *e.g.* Virtual libraries can also be of great help in this regard as it can provide better access to information for people living in isolated areas, Use of electronic blackboards is another medium for gaining access to distance learning and it can also create a virtual classroom which would only prove to be an added advantage.
- ☆ E-learning centres with women trainers and fully developed infrastructure implementing e-learning programmes involving lectures, demonstrations, films, interactions/discussions and question/answer sessions.
- ☆ Using advanced technologies for quality education services: Bringing about new technologies so as to improve the quality of processes like maintaining information, records and other administration-related activities. For *e.g.* making use of online enrolment facility and registering for an education programme is also cost saving.

There are also programmes like Sarva Shiksha Abhiyan (SSA) started by the Government of India to provide quality elementary education including life skills and achieve Universalization of Elementary Education. SSA has a special focus on education of girls' and children with special needs with an aim to provide free and compulsory education for children aged between 6-12 years. It has an agenda of opening schools in area where there is no such facility while strengthening the infrastructure of the existing schools along with a focus on providing quality education. SSA also seeks to provide computer education to bridge the digital divide. There are also other national reforms such as the midday meal scheme, free textbooks, free notebooks, pencils and free school uniforms which proves to be of great help for the girl students.

Women in Science and Technology

The Problem

The number of women in science and technology is alarmingly low rather it is on the decline. It is analyzed and noted that the even after making considerable efforts to increase the participation of women in science and technology and give better access to education in this area, they still remain to be under-represented.

What Could be the Reasons?

- ☆ **Financial Cringe:** Because of the low family income, women drop out from secondary and higher education which is the base foundation for setting up a career in science and technology.
- ☆ **Gender Inequality:** Societal mind set of giving privilege to boys as compared to girls having a strong belief that educating a girl is simply waste of money as they have a whole and sole responsibility of looking after the household

activities some years down the line, as she has to get married and leave. Further, even the teachers in schools do not encourage girls for pursuing science and technology as they again hold a false belief that girls are not intellectually competent enough to study this subject which proves to be one of the major drawback for the girls who possibly want to study further but lack motivation.

- **Gender pay gap:** Women are paid less as compared to men for equal work; they also have less chances of promotion for the same reason. Due to this reason, they get hardly any chance to move up the ladder of their career.
- **Maternity Phase:** Being a woman, she has to plan her family and manage both work and home with incredible efficiency. However, during her childbearing years it gets difficult to meet the high level of work demand and so she has to face a stagnant phase for she cannot expect to grow professionally during that period when she has to compromise for looking after herself and the baby.
- **Lack of Political Will and Women Policy-makers:** There are not sufficient policies for promoting women in science and technology and the absence of women in policy making positions, does not improve this condition.

The Figure 6.4 depicts the percentage of total researchers in Africa, Asia and the Pacific (2007 or latest available year).

It is noted from the data shown that Guinea holds the lowest percentage (5.8 per cent) of female researchers not only in Africa but in all the 118 countries included in the study and only two countries from Africa *i.e.* Lesotho (55.7 per cent) and Cape Verde (52.3 per cent) have so far been able to achieve gender parity. From Asia and Pacific, Myanmar has the highest percentage of female researchers with 85.5 per cent. However, women scientists are very poorly represented in Japan (13 per cent), Bangladesh (14 per cent), India (14.8 per cent), Republic of Korea (14.9 per cent) and Nepal (15 per cent).

ICT Offers a Solution

- ICT helps to create public awareness to women issues providing greater access to technology and information and thus closing the gender gap.
- It provides an access to information and services for women who are culturally or geographically isolated.
- It helps in setting up new businesses and engagement into e-commerce and e-business, to generate income and strengthen existing businesses while creating new employment opportunities.
- It also offers an opportunity to work from home creating more exible working conditions for women.

Income Generation through ICT

Technology needs girls for all sorts of reasons – but perhaps the most important one is that **women drive social and economic growth**: as said by the ITU's Secretary

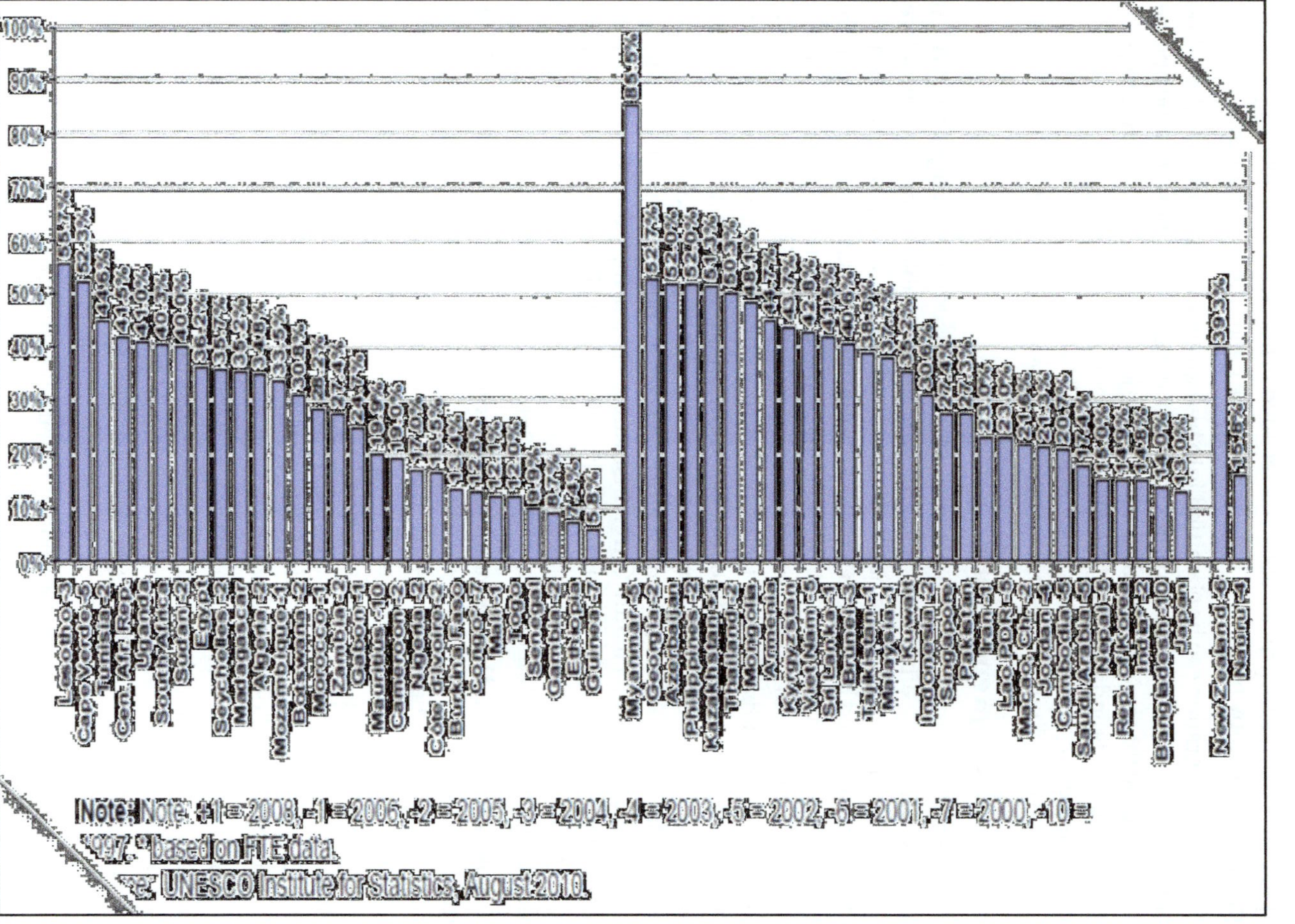

Figure 6.4: UNESCO Statistics Data for the Year 2010.

General Dr. Hamadoun Touré at the "International Girls in ICT Day on 29th May 2012". From last few years, tremendous efforts have been made to empower the women through ICT. Recent advances in information and communication technologies, coupled with increasing privatization has resulted in the creation of new employment opportunities thereby helping to raise the income generation of nations.

Income generation can be classified into two groups, one group including those people who are in search of jobs in order to improve their economic status and another group including those, who provide the jobs to the job seekers with the aim to improve the living standards and increase the capacity of people to produce goods and services that is to generate income. Thus, they are people who help in income generation activities and contribute in the context of overall development of the nation.

Women's Entrepreneurship

"Women entrepreneur" is a woman who organizes and manages any type of enterprise, especially a business, usually with considerable amount of initiatives and risks. Government of India has defined women entrepreneurs as ¯an enterprise owned and controlled by women having a minimum financial interest of 51 per cent of the capital and giving at least 51 per cent of the employment generated in the enterprise to women. In modern society, women entrepreneurial innovation has been recognized as an indispensable factor which contributes as a valuable tool for its economic development and promotion. The knowledge economy has certainly created large number of never before opportunities for women particularly in the service sector. Today, one can see a women entrepreneur in almost every field be it ICT, Retail, Service Sector, Health Care, Insurance, Tourism, Education and even International Trade thus, contributing to the economic well-being of the family and communities, poverty reduction and women's empowerment, thereby helping in achieving the target of Millennium Development Goals (MDGs).

Federation of Indian Women Entrepreneurs (FIWE), a **National Level Organization** who mainly works for the Empowerment of women through ICT initiatives by providing networking platform for women, technical know-how, industry research and expertise, skill development and training while bringing the businesswomen on a Common Forum by organising workshops/seminar on topics such as e-commerce, developing a business plan, etc. Finally, networking knowledge reshapes the women's standard of living and allows them to contribute to the economic, social and political development of the country.

It has been found that in-spite of the various measures taken by the government as well as various national and international organizations, women entrepreneurship in the field of ICT sector is low and it is still considered as a male-dominated field. Women entrepreneurs are facing several obstacles related to their businesses such as:

- ☆ **Lack of Support**: Strong entrepreneurial economy requires the support of wide range of stakeholders which include investors, bankers, customers, suppliers, service providers, family members and others to build a healthy economic environment.

- **National Policy**: There is a lack of clear national policy for promoting women entrepreneurship through ICT.
- **Lack of Skill**: In spite of intensive measures, mostly in developing countries it has been observed that literacy level among women as compared to that in males is very low resulting in inadequate computer skills.
- **Lack of Awareness**: Little awareness of the full range of opportunities offered by ICT other than access to information, limited online information in vernacular language.
- **Technical Problems**: Absence of favorable bandwidth and connectivity for smooth operations.
- **Lack of Networking**: Networks of women entrepreneurs and businesses owners are generally smaller and less diverse than those of their male counterparts.

It is now high time to adopt various means to educate the women about the advantage of ICT and its possibility for their survival as the job creator (entrepreneur) and not only as a job seeker. There are various means adopted in the modern society such as e-commerce to encourage them to participate in ICT sector and create their own individual identity. Besides these there should be some other measures which must be enforced to overcome the problems such as:

- Increase access and use of ICT by organizing national and international workshops and training programmes
- Link women through modern technology and networking to help coordinate agenda, speed up communication and dissemination of knowledge and experiences
- Content development on the web production and use of ICT resources in local languages to help local women to know about other women's entrepreneurial activities
- Set up community information centers to provide information about small scale industries
- Institution of scholarship with incentives to promote enrollment of girls and women in ICT programmes.

Employment

The future of the ICT sector is quite exciting. There are unchartered avenues open to creativity, innovation and entirely new ways of working, interacting and learning that should appeal to women and men alike. ICT job opportunities for women in coming future that includes high speed internet, cloud computing, green ICT goods and services and their "smart" applications thus empowering them to discover a new world in a dynamic way. It also provides innovative job opportunities for women such as power grid informatics, digital media, bioengineering, software designing and so on. Women are in high demand for these jobs, but are conspicuously absent from computer systems administration, technical development and decision-making jobs. Women are very few as producers of information, thus with less access than men to the information and networking resources. Naturally, they have fewer possibilities of orienting technology to address their specific needs. Thus, women's employment figures in ICTs in advanced economies has declined and account for less than 20 per cent of ICT specialists in OECD countries. What are the reasons for this? Some of the probable answers can be:

- Low participation rate of women in global ICT sector; in most of the developing countries women account for only 20 per cent IT jobs.
- Poor literacy among women (in spite of intensive measures to promote education), and inadequate computer skills.
- Unaffordable costs of computer hardware and software, maintenance and connectivity.
- Little awareness of the full range of opportunities offered by ICT other than access to information; limited online information in vernacular languages.
- Absence of favourable bandwidth and connectivity for smooth operations.
- Unequal access to training, resulting in absence of women in most ICT job categories.
- ICT sector, always considered as a male-dominated industry, most high-value and high-income jobs in this sector are occupied by men.
- Gender segregation.

These are not insurmountable barriers; neither do we lack resources to overcome these barriers. Growth in employment, however, has not yet led to a parallel increase in jobs for women in the ICT labor market, with a low female to male ratio. Globally, it has been observed that about less than one third of full-time staff in telecommunications industry are women. Further, it has been also observed that in India only 21 per cent of software professional are women, whereas in Europe 6 per cent and in South Africa only 20 per cent. This result clearly shows that, in most of the developing countries about less than 50 per cent of total staff of women are involved in ICT sectors. Although women are making inroads into technical and senior professions and on average, it has been observed that women accounted for 30 per cent of operations technicians, only 15 per cent of managers and a mere 11 per cent of strategy and planning professionals. What are the measures that should be taken to overcome all these problems?

- Women should be encouraged to come forward and participate in the ICT sector, engaging women and girls in ICT sector work is not only the right thing to do from the point of social justice but it is also smart economics.
- Women should be educated and must possess the power of computer literacy. Computer education therefore becomes essential so that they can not only keep abreast with the current affairs but also learn, earn and fulfill their dreams independently.
- Facilitate women's equal access to resources, employment, markets and trade.

- Eliminate occupational segregation and all forms of employment discrimination. Gender balance in high value ICT jobs in both management and on company boards has been proven to improve business performance.
- Vocational training centre and a placement cell for women and adolescent girls should be encouraged by the corporate groups.
- ICT and technical job-oriented courses should be provided to prepare them for future aspects.
- Governments need to place a premium on promoting ICT skills in primary, secondary and higher education.

'Women make up half the world's population, they use technology as much as men, and they are innovative technical thinkers-so if we want the best technology that we can get, we need diversity at the design table.'

[Lucy Sanders, NCWIT CEO http://research.microsoft.com/enus/collaboration/focus/cs/talent_sanders.aspx]

The growing demand for a range of ICT skills around the globe presents a unique window of opportunity to properly position girls and women in the industry and provide them with the tools necessary to succeed.

Women Health Issues

Women's health is a crucial topic which is greatly affected by the way in which they are treated in the society. Currently, women's health issues have attained higher international visibility and renewed political commitment and target programmes which enable the women to lead a healthier life. In-spite of all these effort made, still women have to face many health problems with gender-based health disparities, limited access to education, and increasing poverty levels are making health improvements for women exceedingly difficult. Especially in rural areas where they face many different health issues as compared to those women who live in towns and cities. Some of the common health problems of women in developing countries are Early Marriages, Malnutrition, Anemia, Hormonal disturbances, Breast Cancer, HIV AIDS, etc. It has been observed that, worldwide one woman dies every 90 seconds during child birth. More than 99 per cent of the estimated 536,000 maternal deaths each year occur in the developing world due to less maternity care, physical abuse during pregnancy, being some of the causes. There are about 39 million cases of HIV/AIDs across the world, out of which 23 million cases are in Sub-Saharan Africa and 57 per cent of those AIDS patients are women whereas in India there are five million cases of HIV/AIDS, of which around 20 lakhs are women; causes may be limited resources available to invest in HIV preventions, prostitution leading to sex with multiple partners, trafficking and rapes. Especially in low-income countries, early and unwanted childbearing also leads women to suffer from HIV and other sexually transmitted infections, and pregnancy-related illnesses sometimes leading to death. Forced marriages and child marriages violate the human rights of women and girls, yet they are widely practiced in many countries in Asia, the Middle East and sub-Saharan Africa which result in weak health, malnutrition, hormonal disturbances, etc. Violence against women is associated with sexually transmitted infections such

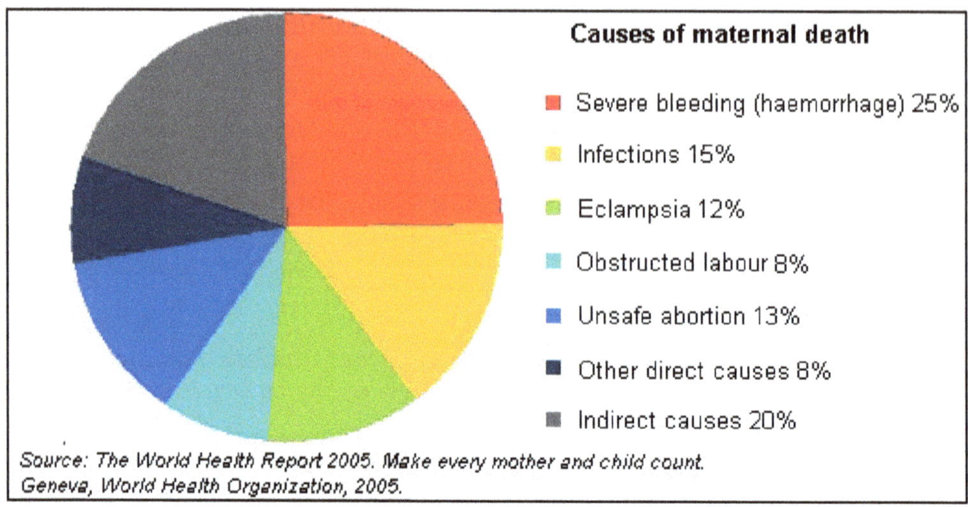

Figure 6.5: Causes of Maternal Death.

as HIV/AIDS, unintended pregnancies, gynecological problems, induced abortions, and adverse pregnancy outcomes, including miscarriage, low birth weight and fetal death.

ICT Solution to Overcome Women's Health Problems

ICT-based initiatives provide medical and humanitarian assistance in emergencies to women in rural areas, *e.g.* in a very small village of Villianur in Pondicherry, India, people are connected through an online database which helps them to access required information which in turn helps the women to get health-related information and all the details about a particular ailment and the name of the doctor who can attend to them. This novel experiment was organized by the M.S. Swaminathan Research Foundation (www.mssrf.org) and has transformed Villianur into the centre of a local area network.

ICT improves the method of dissemination of health care information to remote and underserved areas through a very simple technology called SMS messaging, which has been successfully implemented in Uganda, where birthing attendants are equipped with high frequency radios and are alerted when women need their assistance at home, or when the hospital needs them. This technology works effectively to decrease the maternal mortality.

Facilitate access to the world's medical knowledge and locally-relevant content resources for strengthening women's health research and prevention programmes and promoting women's health, such as content on sexual and reproductive health and sexually transmitted infections, and for diseases that attract full attention of the world including HIV/AIDS, malaria and tuberculosis through m-Health or mobile Health including mobile phones, tablets, PDA, communications satellite, patient monitors, computers, etc.

Tele-monitoring is an application of ICT which involves the method of remotely monitoring patients when the doctor and the patients are not at the same location. Here, patient will have a number of monitoring devices at home, and the results of these devices will be transmitted via telephone to the health care provider or doctor. Thus, it proves to be beneficial for women especially in the developing countries to share their problems with health care providers anywhere in the world and ask for solutions sitting at their homes.

Telemedicine is the use of ICT in order to provide clinical health care at a distance as well as the transmission of medical, imaging and health informatics data from one site to another, thereby eliminating distance barriers and improving an access to the medical services that would often not be consistently available in distant rural communities. Thus, it also helps in improving the condition of women especially in isolated communities and remote regions by imparting necessary information and proper health care solution.

Promote the development of international standards for the exchange of health data through Electronic Patient Records (EPR) and thereby help the Professionals to communicate the patient data between different Health Care Professionals or Doctors across service providers.

Alert, monitor and control the spread of communicable diseases, through the improvement of common information systems including multimedia technology and wireless health care delivery systems.

Female Infanticide and Female Foeticide

India's legal framework stipulates equal rights for all, regardless of gender. In practice, however, unequal power equations between males and females have led to violations of women's reproductive rights.

Female infanticide is a deliberate and intentional act of killing a female child within one year of its birth either directly by using poisonous organic and inorganic chemicals or indirectly by deliberate neglect to feed the infant by either one of the parents or other family members. On the other hand female foeticide is the termination of the life of a foetus within the womb on the grounds that it's sex is female and is also known as sex selective abortion.

The girl child has often been a victim to the worst forms of discrimination. Gender bias, deep-rooted prejudices, and discrimination against the girl child have led to many cases of female feticide in the country.

Female infanticide now in most places has been replaced by female foeticide. Denial to a girl child of her right to live is one of the heinous violations of the right to life. It has, however, been reported that the moral guilt attached to elimination of the girl child after she is born is not felt equally if the child is eliminated while still in the womb. Hence female infanticide is getting replaced more often by female feticide.

Female infanticide cuts across all social and economic boundaries. In rural areas the lack of education, economic resources, and access to healthcare are major factors that lead to the massacre of infant girls. In urban areas, selective abortion is commonly employed by individuals with access to modern medical technology that allows for early detection of sex. Unfortunately, although Government programmes and human rights organizations strive to put an end to these practices, female infanticide continues.

Women are under tremendous pressure to give birth to male babies. In most cases reported, they are named as the murderers of female children. It is either the mother-in-law or the mother or the midwife who commits the murder. We need to ask whether these women are violators or victims. Even though it is the father who determines the sex of a child, the mother is condemned by the society as the one unable to give birth to a son!

Many women themselves are interested in knowing the sex of the unborn child and they do not see any moral problem in undergoing these tests. Secondly, most women have an inherent son complex. They know for certain that their status in the eyes of their family, extended family, community and the village as a whole will go up with the arrival of a son. Gifts will flow in, there will be celebrations and relatives from far and near will visit and call on them. On the other hand, if there is a daughter there is general gloom, no celebrations, no gifts and the image of the woman suffers badly.

Incidence and Magnitude

"During 2001- 2011, the share of children to total population has declined and the decline was sharper for female children than male children in the age group 0–6 years," said the study "Children in India 2012- A Statistical Appraisal" conducted by the Central Statistical Organisation, India.

"Though, the overall sex ratio of the country is showing a trend of improvement, the child sex ratio is showing a declining trend, which is a matter of concern," the study said.

According to the report, female child population in the age group of 0-6 years was 78.83 million in 2001 which declined to 75.84 million in 2011.

The population of girl child was 15.88 per cent of the total female population of 496.5 million in 2001, which declined to 12.9 per cent of total number of 586.47 million women in 2011.

In India the most dramatic drop in the child sex ratio has taken place in the states of Punjab, Haryana, Himachal Pradesh, Gujarat and Maharashtra where clinics specializing in sex determination and sex-selective abortions are known to have been in existence for at least a couple of decades.

The child sex ratio has dropped from 945 girls (0 – 6 yrs) per 1000 boys (0 – 6 yrs) in 1991 to 927 girls per 1000 boys in 2001 to 914 girls per 1000 boys in 2011 (Table 6.1). Female foeticide in 21st century India is a biggest challenge against the laws of the land in general and women's empowerment in particular. Eradication of this practice is the most urgent need of the hour and thus becomes a genuine concern of each one of us. Table 6.2 shows the Incidence and Percentage Contribution of crimes committed against children during 2000 (State and UT-Wise) and the Table 6.3 shows incidents of infanticide and foeticide during 2007-2011 in India.

Table 6.1: Sex Ratio (Females per 1000 males) and Child Sex Ratio (Girls per 100 boys in the age 0–6 yrs), India, 1961–2011

Year	*Sex Ratio*	*Child Sex Ratio*
1961	941	976
1971	930	964
1981	934	962
1991	929	945
2001	933	927
2011	940	914

Source: www.academia.edu/./Tables_on_Sex_Ratios_in_India_and_World.

Table 6.2: Incidence (I) and Percentage Contribution to All India (P) Crimes Committed Against Children during 2000 (State and UT-Wise)

Sl.No.	*States*	*Foeticide*		*Infanticide*	
		I	*P*	*I*	*P*
1.	Andhra Pradesh	8	8.8	7	7.7
2.	Assam	0	0.0	4	3.8
3.	Bihar	1	1.1	4	3.8
4.	Gujarat	0	0.0	4	3.8
5.	Haryana	13	14.3	1	1.0
6.	J&K	0	0.0	1	1.0
7.	Karnataka	1	1.1	2	1.9
8.	Kerala	0	0.0	2	1.9
9.	Madhya Pradesh	14	15.4	31	29.8
10.	Maharashtra	41	45.1	20	19.2
11.	Orissa	1	1.1	0	0.0
12.	Punjab	0	0.0	6	5.8
13.	Rajasthan	9	9.9	5	4.8
14.	Sikkim	0	0.0	3	2.9
15.	Tamil Nadu	0	0.0	8	7.7
16.	West Bengal	0	0.0	2	1.9
17.	Chandigarh	1	1.1	0	0.0
18.	Delhi	2	2.2	2	1.9

Source: Crime in India 2000, p. 216.

Diaz, (1988) states that in a well-known Abortion centre in Mumbai, after undertaking the sex determination tests, out of the 15,914 abortions performed during 1984-85 almost 100 per cent were those of girl foetuses. Similarly, a survey report of women's centre in Mumbai found that out of 8,000 foetuses aborted in six city hospitals

7,999 foetuses were of girls (Gangrade, 1988: 63-70). It is reported that about 4,000 female babies are aborted in Tamil Nadu (southern India) every year. Sex determination tests are widely resorted to even in the remotest rural areas. Since most deliveries in rural areas take place at home there is no record of the exact number of births/deaths that take place. Therefore, it is difficult to assess the magnitude of the problem in the rural and geographically remote areas.

Table 6.3: Incidents of Infanticide and Foeticide during 2007-2011

Sl.No.	*Crime Head*	*Year*					*Percentage Variation in 2011 Over 2010*
		2007	*2008*	*2009*	*2010*	*2011*	
1.	Infanticide	134	140	63	100	63	–37.0
2.	Foeticide	96	73	123	111	132	18.9

Source: Ministry of Home Affairs, annual report 2012-13.

Factors Leading To Female Foeticide

In India female foeticide is taking place for various reasons *viz.* economic, socio-ritual, and technological.

a) Economic Factors

The female foeticide in the 21st century have a great deal to do with capitalist modernity. There are aspects of it lying behind this phenomenon.

1. For rural households with landed property there is a clear inverse correlation between the income level and child sex ratio. It is especially evident in south India. Again there is gender based wage level. For the same work females are paid less remuneration. In most cases women enter in the domestic non-paid services which a patriarchal society gives little or no value at all, so they are regarded as liability than assets.
2. Cultural politics of dowry in the Indian society have a lot of answer for this pernicious phenomenon. Since the turn of century the recorded dowry deaths are increasing. Nearly 7-8000 brides are murdered per year for the lack of full payment of dowry. Nearly 3-5000 brides are committing suicides for dowry related harassment. Brides are thought as commodities and the pre-marriage and marriage have been described as 'consumption oriented reproductive journey'. When the reproductive practices make daughters into such economic burden, the threat of having to amass dowry is motive enough to dispose female commodities (Barbara Harriss-White, 2009).
3. The female foeticide has been commodified. It has started to become a field of accumulation in its own right. Malini Bhattacharya, the member of the national commission for women, admitted that in the era of liberalization "one has to allow freedom of choice to the service seeker and the freedom to sell by the service provider". Foeticide may cost one or two month's earnings, while dowry requires mobilization of several years' income. Hence there appears equilibrium between service seeker and provider. UNICEF

estimates that the turnover of foeticide industry has now reached 244 million dollar from 77 million dollar in 2006 (Barbara Harriss-White, 2009). Those who disapproved of the practice of sex selective abortions but engaged in it against their principles expressed their compulsions and helplessness due to pressures arising out of unhealthy competition in the health care service sector. It was said that if they did not provide abortion care services, some others would have provided them (Tandon and Sharma, 2008).

For these economic reasons females are not desired. Here we can quote an old folk song relevant in this context.

'Oh, God, I beg of you,
I touch your feet time and again,
Next birth don't give me a daughter,
Give me Hell instead.'

—An old Folk Song from Uttar Pradesh

b) Socio-ritual Factors

Females are vulnerable to brutalities of the male in the forms of physical, mental and sexual assaults and traumas in the patriarchal societal structure of India. Females are subjugated, condemned, and deprived in all spheres of life. Every parent of a girl child is at risk for their daughter in this patriarchal society for the above-mentioned causes. Again for the funeral ceremonies of the parents, presence of a son is a must. According to Manu, A man cannot attain moksha **(redemption)** unless he has a son to light his funeral pyre. It is common to believe that the sons will take care of them in their old age. These socio-rituals factors including illiteracy and orthodox society norms lead to craving for a male baby, discarding the females one after another, year after year.

c) Technological Factors

The presence of low-cost technologies like ultrasound, have led to sex-based abortion of female foetuses, and an increasingly smaller percentage of girls are being born each year (Jain, 2005).

d) Population Policy

Indian family planning policies promote a two-child family and health workers say this often leads to abortion of female foetuses in efforts to have a "complete family" with at least one son (Sen, 2005).

e)

The increase in rate of female foeticide is a result of the greed and unethical practices of the medical community.

f)

The enforcement of the laws against female foeticide is poor, with a very low rate of prosecution of offenders, including the medical practioners, and extremely poor conviction rate.

g)

Practice of infanticide had been present in a few communities in some districts and regions in India but this practice did not reach the alarming proportions in elimination of girls as the present day availability of sex determination followed by sex selective abortions.

This is also because in foeticide, there is no inhibition from actually killing a child that an act of infanticide would involve.

ICT Initiatives to Combat Female Foeticide/Infanticide in India

Information and Communication Technologies [ICTs] are a diverse set of technological tools and resource to create, disseminate, store, bring value addition and manage information. The ICT sector consists of segments as diverse as telecommunications, television and radio broadcasting, computer hardware, software and services and electronic media, for example, the internet and electronic mail. ICTs are emerging as a powerful tool for gender empowerment in a developing country like India. There has been a rapid growth in the ICT sector since the late 1980s and the use of ICT has dramatically expanded since the 1990s.

The Ujjas Innovation: The National Foundation of India, a non-profit foundation initially offered the village women from the Western state of Gujarat's underdeveloped region called Kutch to bring out their own newsletter called Ujjas (which means the 'LIGHT') with the help of Kutch Mahila Vikas Sangathan, a district level NGO. The newsletter was very successful amongst the women to help fight social exploitation as well as issues such as dowry, female infanticide, drunkenness amongst the men folk, and enabled them to trade and do business amongst themselves as well as share knowledge amongst them. The success of Ujjas attracted other funding agencies including the Ministry of Rural Development to support a 105-episode community radio programme also called the UJJAS that is broadcasted by the All India Radio Bhuj station. Radio Ujjas is a successful community radio initiative that works to sensitize the people of Kutch on local issues, particularly on matters related to women. The programme allows the women to voice their concerns, learn from each other and interact with the rest of the world. The impact of Ujjas in one of the remotest corners of India is a testimony to the fundamental belief that the innovative use of communication technologies can be a powerful tool in the hands of the poor, particularly women and the children. There is a similar programme in **Nepal** for creating community awareness known as **Radio Sagarmatha.**

The Health and Family Welfare Department, Govt. of India as well as media and related agencies have managed to create a high degree of awareness as well as concern over female foeticide in the country. ICTs apart from sensitizing people against this heinous crime can also play a highly interventionist role by proactively pursuing cases against erring doctors, booking them under the law of the land as well as helping people in general to change their opinion about "a girl child".

Datamation Foundation acutely conscious of this monumental crisis decided to take the initiative on its own to implement an ICT enabled strategy for saving the girl child. A portal was established apart from a series of path-breaking steps to advocate the rights of the girl child to live apart from spreading the message of greater gender equity were implemented. These steps ranged from running SMS campaign with the help of private carriers, email campaigns in the areas most affected by the population crisis. The Foundation also made a 10 second TV spot on the selective sex abortions and this film is being broadcasted on a number of TV channels. The film went on to get an award at the Cannes Film Festival. More importantly they implemented an online e-Governance module for lodging complaint against doctors/radiologists doing illegal sex abortions. To the users for facilitating easy complaint lodging; online databases of registered ultra-sound machines and competent authorities mandated to take punitive action against all those who perform illegal sex tests where provided as well.

'Save the Girl Child Campaign', which uses ICTs to generate and record complaints against members of the medical community indulging in selective sex determination tests and selective abortion of female fetuses.

"Silent Observer" (shown in Figure 6.6) is a special tamper proof device helping Maharashtra government identify cases of female foeticide. Information provided by this device, which is fitted to a sonography machine, is vital not only to track pregnancy tests but also to detect unreported terminations.

Figure 6.6: Silent Observer.

With the skewed child sex ratio in the state showing few signs of improvement and female foeticide on the rise in rural areas as well, Maharashtra has turned to a gadget called the Silent Observer for help. Once fitted in sonography machines, this device maintains a log of all pregnancy tests done in a year, helping track under-reporting or false reporting of pregnancy termination cases.

Girish Lad, CEO of Magnum Opus which helped fit the device, said it takes video inputs from the personal computer and stores them in a compressed format in a local drive - in a preconfigured video or image format. It blocks all other ports except those required for data input to ensure additional data security in the device, said Lad.

"This machine will enable the administration to provide images in black and white and use as evidence in case there has been termination of pregnancy," said Lad.

The results have been encouraging. According to advocate Salunkhe, an Anganwadi survey done in the district showed that the child sex ratio in Kolhapur had climbed to 854 females per 1,000 males (earlier it was 795 females per 1,000 males). Health authorities in Haryana will also install silent observer chips in all ultrasound machines to monitor the ultrasound processes being done by doctors at private clinics.

'Fight-Back', an online gender justice campaign is mobilizing Indian medicos on their web-based platform to spread awareness on the shocking practice of female foeticide.

Fight-Back, which was set up in 2008, has declared March 8 as the International Missing Women's Day to highlight the missing number of girl-children who fall victims to foeticide while they are in the womb.

Fight-Back has brought 130 doctors on one platform to "condemn female foeticide" and to "issue a pledge" on March 8 that they will not abet illegal sex determination tests and aborting of female foetuses.

Media-both print and electronic-plays a very significant role in removing gender bias and developing a positive image of the girl child in the society, but in a county like ours where there are problems in reaching the backward, rural and tribal areas, a mix of mass media with various traditional forms of communication may provide a more effective alternative to influence the illiterate and the poor.

Awareness Campaigns- To deal with a problem that has roots in social behavior and prejudice, mere legislation is not enough. Various activities have been undertaken to create awareness against the practice of prenatal determination of sex and female foeticide. To implement the provisions of the Act the help of media units like AIR, Doordarshan, Song and Drama Division, Directorate of Field Publicity, Press Information Bureau, Films Division and Directorate of advertising and visual publicity (DAVP) is also being sought.

The TV soap opera, 'Atmajaa' (Born from the Soul), is an innovative new approach designed to shock the nation out of its obsession with gender selection. The 13-part series, telecast at primetime on national television, looks at the laws surrounding prenatal diagnostic tests, gender, poverty, anti-dowry laws, violence against women and the problems that occur when there are too few women. The series raises awareness about the use of prenatal gender screening, despite it being banned in India since 1994. It also explores the reasons why people from all classes continue to prefer sons to daughters.

e-governance modules can help identify ultrasound clinics, nursing homes carrying out illegal sex determination tests. Complete database of the ultrasound machines, nursing homes is available online at the websites. The complaints can be lodged by anyone against doctors performing illegal sex tests. Once a complaint gets filed, the competent authority takes over the investigation of the complaint automatically and attempts to bring the culprits to book under "The Pre-Natal Diagnostic Techniques (PNDT) Act".

Violence Against Women's Safety and Security: Violence against Women

The international community recognizes 'violence against women' as any act of gender-based violence that results in, or is likely to result in, physical, sexual or psychological harm or suffering to women. Gender-based violence is violence that is directed against a woman because she is a woman or affects women disproportionately. Examples include rape, domestic violence, trafficking, forced prostitution, sexual exploitation, sexual harassment, female genital mutilation and forced marriage. Violence against women includes threats of such acts, coercion or arbitrary deprivation of liberty. Violence against women can occur in private (such as in the home) or in public settings (including places of work and educational institutions).

Women live in a very insecure world indeed. Many fall victim to gender selective abortion and infanticide (boys being preferred to girls). Others do not receive the same amount of food and medical attention as their brothers, fathers and husbands. Yet others fall prey to sexual offenders, to 'honour killings' and to acid attacks (most often for refusing a suitor). An estimated 5,000 women are burnt to death each year in 'kitchen accidents' because their dowry was seen as being too modest. Scores succumb to the special horrors and hardships that conflict, war and post-conflict situations reserve for girls and women. A shocking number of women are killed within their own walls through domestic violence. Rape and sexual exploitation remain, moreover, a reality for countless women; millions are trafficked; some sold like cattle.

In South Africa a frightening 40 per cent of girls aged 17 or under are reported to have been the victim of rape or attempted rape. According to the United Nations Children's Fund (UNICEF), 100 to 130 million women around the world have been genitally mutilated. This would translate to a figure as large as 2 million girls or more being genitally mutilated each year.

However with the rise of new media, advances in technology and increased access, new opportunities exist in which violence can be further prevented. One of the most popular forms of programme implementation using mobile phones is through the utilization of **short message service (SMS) messaging** or better known as text messages. SMS has been used most frequently for educational awareness via one-way messaging.

Use of SMS and geocoding can be applied to the prevention of interpersonal violence as well as collective violence. In Egypt, Frontline SMS, a text messaging based system, in conjunction with the Ushahidi platform have been used in a project called **Harassmap** (Figure 6.7). In Egypt 83 percent of women are exposed to sexual harassment. The basis of the programme is that if a woman is sexually harassed she can send an SMS to the Harassmap number with corresponding details of the incident. This information will then be mapped on the website, allowing for "hot-spots" of harassment to become identified. They also provide help and information for victims. This programme aims to help break through the silence that surrounds this issue.

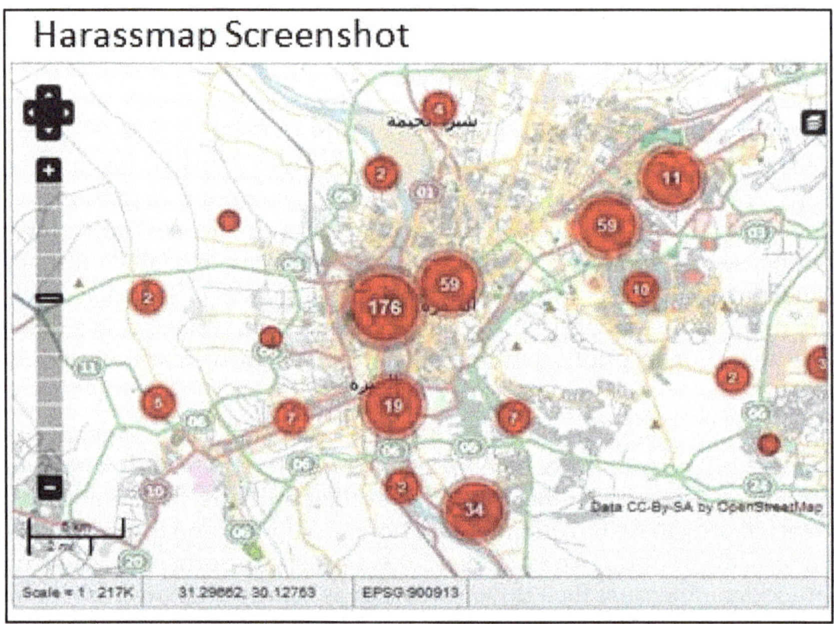

Figure 6.7: Screenshot Showing Harrasmap.

Another programme aiming to break through some cultural barriers is the **Mobile Cinema Foundation**. This programme travels to various soldiers' camps and exposes soldiers to the consequences of rape through short films. These films seek to educate the soldiers of Congo's National Army through victim's testimonies and discussions that are held after the viewing. This form of digital storytelling can be a powerful tool both for empowering the victims and educating offenders on the effects their actions can elicit.

Again conventional forms of technology and media can play a role in the prevention of intimate partner violence. This includes hotlines and awareness campaigns through both traditional and new media. **Bell Bajao, or Ring the Bell, campaign,** launched in India in 2008 is a series of public service announcements urging men and boys to stand against domestic violence. The idea is that if one is hearing violence in progress to ring the bell and ask a simple question, such as "Can I borrow a cup of sugar?" It is likely the perpetrator recognizes the person has heard the violence and this will interrupt the action**. The organization also hosts a blog** for victims to voice their experiences and offers information and guidance.

In 2010, Breakthrough's video vans travelled 14,000 miles through cities and villages screening these PSAs and involving communities through games, street theatre and other cultural tools resulting in a sustainable, on-ground process of transforming hearts and minds.

Bell Bajao's tools and messages have been adapted by individuals and organizations around the world, including Canada, China, Pakistan and Vietnam.

The campaign has won Breakthrough 23 awards including the Silver Lion at the 2010 advertising festival held at Cannes.

Women's rights organizations are using social networking tools to support women whose lives are threatened, to take action that prevent violence, and seek redress for women and girls. Women's rights activists are adapting **GPS applications** in mobile phones – intended for commercial purposes – to warn others of dangerous areas and document abuse. In South Africa, **Women'sNet** is teaching girls to avoid harassment through cell phones and "Keep your chats exactly that!" is a campaign by Girl'sNet, a daughter project of Women'sNet, designed to empower young people to prevent them from becoming victims of harassment, bullying and violence when using the internet and cell phones. It also works to encourage strategies for using ICTs in affirmative ways to advocate for change on issues that concern young people.

Survivors of violence are producing digital stories to denounce what happened to them in their own words and voices and at the same time connect with others, build solidarity and aid the healing process.

Some women's groups in Thailand have started to use the Internet in their work. An example is the **Development and Education Programme for Daughters and Communities Centre (DEPDC)**, whose website can be found at www.depdc.org. DEPDC became internationally famous when its founder and manager, Sompop Jantraka, was named an Asian Hero by *Time* magazine in April 2002. Its main objective is to prevent young women, from seven districts in the northern Thai province of Chiangrai, from entering the sex industry or the labour market at a young age.

There are similar examples of innovative use of the Internet in Pakistan as well. The project **'Portrayal of Women in the Media'** was designed by UNDP in collaboration with Pakistan Television (PTV), Pakistan's only public telecast media channel. It aims to ensure a positive, balanced and diverse portrayal of women in electronic media through training and gender sensitization of TV media professionals.

Under this programme, PTV facilitates specialized courses on gender sensitization and seeks to incorporate gender issues into regular training programmes. Gender committees have been established at all PTV centres and, in 1999, PTV management made a major policy decision to ban all displays of physical violence against women and children in their programmes. Perhaps the most significant landmark in the history of PTV's gender policy was the launching of a daily 60-minute transmission for women, titled *Khwateen Time*. The project has now established its credibility by securing reasonable success in portraying a better image of women through the electronic media. Mind sets and attitudes cannot be changed overnight but at least the first step has been taken in that direction.

Shirkat Gah (Urdu for "convening point for participation") is one of Pakistan's most respected women's rights groups, known for its efforts to empower women and advocate for a better legal and social framework. In recent years, the international recognition it has received, and the coalitions and networks it has helped build, have served the cause of Pakistani women through the Internet and email. Shirkat Gah has been able to leverage the immense reach of the Internet and email to help draw national and international attention to some of the most disturbing recorded cases of rape and violence against women.

The APC (**Association for Progressive Communications), South Africa** launched a global campaign **"Take Back the Tech!"** to end violence against women which is growing in strength and numbers with people from various developing countries taking part.

With the spate of rape cases being highlighted across India especially in the backdrop of the heinous gang rape case in Delhi there has been renewed concern over the safety of women like never before. **These apps help alert your near and dear ones, and tell them about your exact location in case of emergencies:**

1. FightBack

Availability: FightBack is available for select Android, Blackberry and Nokia smartphones. Users can go to www.fightbackmobile.com, login with their Facebook ID and follow on-screen instructions to download the app for their handset.

FightBack is an India specific application available for all types of mobile phones. FightBack uses GPS, SMS, location maps, GPRS, email and your Facebook account to inform your loved ones in case you are in danger. You can add up to five emergency contacts to the list. When the 'Panic' button is pressed from the mobile application, the portal alerts page gets updated with the live alert data and shows the location of the alert on the Google Map.

This will be visible to all the users of the web portal. The portal also sends out SMS messages to the mobile numbers pre-set by the user. When this hyperlink is clicked, it will show the location of the app user on Google Maps. The web portal also updates the user's Facebook status with the SOS message. This SOS message will be visible to all the friends who are connected to the user via Facebook.

When the Facebook message is clicked, it will take the Facebook friend to the web portal Alert Page, and will show the location of the mobile user when the SOS was raised, along with a time stamp. This app is a paid application.

2. Safe

Availability**:** Android, iOS and BlackBerry phones and tablets.

This is a GPS based safety alarm that sends SOS messages during emergencies to the mobile numbers you have previously chosen. All you have to do is select a few numbers of your relatives and friends. You can add as many numbers as you want. It comes with a big red button for SMS/call in case of emergencies. You just need to push that, triggering the app to send instant SOS messages to the numbers you have chosen.

3. Life 360 Family Locator

Availability: Android as well as non-smartphones.

This too uses GPS tracking technology. With this app, you can locate family members at any time using GPS, Wifi and Cell Triangulation technology.

However, when it comes to security, this is one of the best apps. When you press the panic button, the app sends an SOS through email, SMS or app notification if the other family member has the app as well.

4. Circle of 6

Availability: Requires iOS 4.0 or later and is compatible with iPhone 3G, iPhone 3GS, iPhone 4, iPhone 4S, and iPhone 5.

As its name suggests, this app lets you choose six trusted friends to add to your circle. If you get into an uncomfortable or risky situation, use Circle of 6 to automatically send a pre-programmed SMS alert message with your exact location to your circle of friends. The app's usefulness is verified from the fact that it was the winner of the 2011 White House 'Apps against Abuse' Technology Challenge.

5. SOS Whistle

Availability: All phones

SOS Whistle does not send an SOS or let your friends know where you are. Rather, as its name suggests, the app has a very simple function-to trigger a whistle. In fact, it can trigger a whistle sound even if the phone is in silent mode. Thus, it is a great app in times of danger to tell other passers-by of your situation. You don't even need a data connection or GPS service. You just have to tap the screen and the app will start an alarm.

Keltron, Kerala State Electronics Development Corporation Limited is planning to launch cutting edge technology solutions for women safety in public transport, which aims at promoting the 'women-safe vehicle concept'.

The product launched in April 2013 includes solutions that would cover in-vehicle surveillance, vehicle and passenger tracking, passenger emergency call and emergency assistance.

Keltron intends to use a combination of world class technologies and their in-house manufacturing strength to deliver quality safety solutions for not only women community but also to all vehicle users and owners.

Kapil Sibal, India's *Minister of Communications and Information Technology*, has expressed his wish for a **GPS enabled wristwatch** to ensure women's safety. The minister wishes for the watch to have features such as GPS, video recording, alarm and capable of calling up dedicated lines – and all that packed into a price less than Rs. 1,000. While the idea behind it seems definitely noble, it remains to be seen whether it would be properly executed and see the light of day.

Upliftment of Rural Women

Empowering Women through Self-Help Groups (SHGs)

The Government of India and state authorities alike have increasingly realized the importance of devoting attention to the economic betterment and development of rural women in India. The Indian Constitution guarantees that there shall be no discrimination on the grounds of gender. In reality, however, rural women have harder lives and are often discriminated against with regard to land and property rights and in access to medical facilities and rural finance. Women undertake the more onerous tasks involved in the day-to-day running of households, including the collection of fuel-wood for cooking and fetching of drinking water, while their nutritional status and literacy rates are lower than those of men. They also command lower wages as labour: as rural non-agricultural labourers, women earn 44 rupees per day compared to 67 rupees for men. Women's voice in key institutions concerned with decision making is also limited. In 2007, only 8 per cent of all seats in the national parliament were occupied by women.

Self-Help Groups

Self-help groups are generally facilitated by NGOs, and increasingly advise and train members in a variety of on- and off-farm income-generating activities. Indeed, in a number of recent projects, NGOs were substituted by trained facilitators and animators drawn from self-help groups. Through promotion of self-help groups, IFAD-funded projects have contributed to improving the overall status of women in terms of income, empowerment, welfare, etc.

India Launched Sanchar Shakti to e-mpower Rural Women

In an attempt to bridge the digital divide, the Sanchar Shakti scheme envisages creating ICT skills among rural women by providing useful information to women about health, social issues and government schemes over their mobile phones. The scheme seeks involvement of women Self Help Groups for its successful implementation.

Former President Pratibha Patil launched DoT-USOFs' pilot project Sanchar Shakti, which aims to provide useful information to women, about health, social issues and government schemes over their mobile phones.

The scheme, which is funded by Department of Telecom's (DoT) Universal Service Obligation Funds (USOFs), is for mobile value-added services and ICT related livelihood skills for Women's Self Help Groups.

The scheme is initiated as a gender budget scheme; it adopts an innovative approach to connect rural women with the ICT sector, by involving women Self Help Groups (SHGs).

It will also serve the purpose of increasing the teledensity, as well as broadband connectivity in rural areas, enabling them to join the cyber community and in this way, bridge the digital divide, as stated in the details of the scheme.

With the creation of ICT skills sets in rural areas, over a period of time, these areas can become centers for the consumption and production of various ICT goods and services, including hubs for processing outsourced work, it added.

References

1. Annan, K. (2002). Information and communication development: Information society surmit. p. 7.
2. Bruce, O. (1995). Internet with a difference: Getting people hooked up. Available: www.zwren.org.zw/publications/information

3. Crede, A., and Mansell, R. (1998). Knowledge societies. in a nutshell: Information technologies for sustainable development. Ottawa, Canada: IDRC.
4. Global Monitoring Report (GMR) 2006, 'DFID's girls' education strategy Girls' education: towards a better future for all', p. 35.
5. Jean-Yves Hamel, August 2009, 'Information and Communication Technologies and Migration', United Nations Development Programme, Human Development Reports, Research Paper 2009/39, pp. 1-37.
6. Shen,K., Zhang, P. and Tarmizi, H. 2009, 'The role of culture in women interacting with ICT', International Conference on Information Resources Management, pp. 1-3.
7. Stella N.I. Anasi, PNLA Quaterly 76:3 (Spring 2012), 'Information and Communication Technologies and Women's Education in Nigeria: Challenges and Future Directions'. Available: http://unllib.unl.edu/LPP/PNLA per cent 20Quarterly/anasi76-3.pdf.
8. Thioune, R.M.C. (2003). Information and communication technologies for development in Africa: Opportunities and challenges for community development. Volume1. Ottawa: IDRC. Available: http://www.idrc.ca.
9. http://www.itu.int/en/ITU-D/Statistics/Documents/facts/ICTFactsFigures2013.pdf.
10. http://www.issa.int/News-Events/News2/UN-International-Women-s-Day-2013-Bridging-the-gap.
11. http://ssa.nic.in.
12. http://www.scidev.net/global/education/feature/overcoming-gender-barriers-in-science-facts-and-figures-1.html.
13. http://www.sciencedaily.com/releases/2012/10/121003082719.htm.
14. http://www.uis.unesco.org/ScienceTechnology/Documents/sti-women-in-science-en.pdf.
15. http://wcd.nic.in/research/ict-reporttn.pdf.
16. http://www.uis.unesco.org/ScienceTechnology/Documents/unesco-egm-science-tech-gender-2010-en.pdf.
17. www.depdc.org.
18. Federation of Indian Women Entrepreneurs; www.gemconsortium.org.
19. http://smallb.in/per cent 20/fund-your-business per cent 20/additional-benefits-msmes per cent 20/women-entrepreneurship.
20. http://www.nordregio.se/Global/Research/1437_Paper_Gender per cent 20Work per cent 20and per cent 20Organization_Keele.pdf.
21. http://www.indiatogether.org/women/health/health.htm.
22. http://www.mssrf.org.

Part II

Important Role of NGOs in Women Empowerment through ICT with Special Focus on Rural Women

Chapter 7

ICT Framework Used as Women's Tool for Poverty Alleviation in Rural Areas of Kenya

Catherine Kathure Kaimenyi

Assistant Lecturer, Chuka University,
Chuka, Kenya
E-mail: ckaimenyi@yahoo.com

ABSTRACT

This paper presents a study that investigated the information and communication technology (ICT) framework used in rural areas of Kenya. The study uses a gender perspective to establish the link between poverty and ICT. Although women play an important role in socio-economic development of any society, but majority of women in rural Kenya are afflicted with poverty, ignorance and disease. While it is recognized that there are numerous factors responsible for the poor condition of women, this research believes that lack of ICT is one of the major contributing factors. This observation stands on the basis that much literature and common belief support the use of ICT in poverty alleviation. Although both the government and the private sector have put unending efforts to improve access to ICT, there is still no ICT framework used in poverty alleviation in Kenya.

The study utilized secondary data from: researchers in the field of gender and poverty alleviation, poverty data from the Kenya National Bureau of Statistics, the National ICT status data from Communications Commission of Kenya, and Vision 2030 projections, to develop a framework of ICT used by rural women. Critical questions raised in the study, include: (i) What are the poverty patterns in Kenya? (ii) What application of ICTs are available for women in Kenya? (iii) What are the ICTs growth patterns in Kenya? (iv) What is the relationship between poverty and ICT? (v) What framework should be put in placing ICT use by rural women in poverty alleviation in Kenya? The results show that poverty is related to a gender dimension, leaving women more vulnerable because of their roles in the society. The study recommends that poverty needs to be tackled effectively through training and education with special focus to women, provision of information and creation of supportive infrastructure coupled with policy formulation which supports their affordability and sustainability.

Keywords: *Women, Rural, Kenya ICT Poverty-alleviation, Framework.*

Introduction

Women play an important role in socio-economic development of any society because of their primary role in the family unit. However, as revealed by the United Nations (2006), women are more likely to be affected by poverty than men, because of their unequal access to economic opportunities. In addition, in general women suffers from lack of resources including credit facilities, land ownership, lack of access to education and cultural practices, minimal participation in decision making processes, poorly distributed infrastructure, poor health facilities, and high levels of HIV/AIDS infections.

In Kenya, poverty manifests itself in the forms of hunger, malnutrition, illiteracy, lack of shelter and failure to access essential social services such as basic education, health, water and sanitation. This is observed in Kenya Vision 2030, the current Blue-Print that is guiding the Government Development Agenda, up to 2030. In the research on gender and poverty reduction in Kenya, Kimani E.N., Donald K. (2010) noted that majority of the poor Kenyans are women, as they are deficient in access to educational opportunities, due to the low value stated on the girl child, as compared to the boy. Based on the traditional believes and practices, women have less or no ownership, access and control to family assets and resources, as compared to their male counterparts. In this respect, in the incidences of deprivation through poverty, they are more vulnerable. Unfortunate deprivation for women has serious negative impact to the children, because of the reduced performance in both reproductive and productive roles. On the other hand, in the Kenyan social context, children stand to benefit more directly when women are well endowed as most of the resources are used at the household level.

This paper seeks to understand patterns and trends of poverty and information and communication technology (ICT) development in Kenya with special focus on gender and rural areas. A preposition is made that there cannot be sustainable development aimed at poverty reduction, without mainstreaming women in all development structures and processes. This culminates with formulation and implementation of gender responsive policies.

Objectives

The specific objectives of the study were to:

1. Understand the poverty patterns in Kenya
2. Establish the availability of ICTs for women in Kenya

3. Establish the ICTs growth patterns in Kenya
4. Establish the relationship between women, poverty and ICT
5. Propose a framework for use of ICT in poverty alleviation in rural areas

Materials and Methods

Study Design

The study was carried out from a qualitative paradigm using secondary data and involved the summary, synthesis and analysis of existing research. Most of the data was collected from the Kenya National Bureau of Statistics (KNBS – poverty data), the Communications Commission of Kenya (CCK –ICT status data), as well as documented studies in the field of gender, poverty and ICT.

Defining Poverty

People have different perceptions and definitions of poverty globally (Robeyns, 2005). Poverty is often defined in terms of individual's minimum purchasing power set at a certain amount. According to Alkire (2005), poverty is multidimensional and takes the shape of hunger and inadequate nutrition, slum housing or homelessness, unhygienic living conditions, child and maternal mortality, unsafe water and ragged clothes among other distinctiveness. The United Nations Development Programme (2008) categorizes poverty as income poverty, capability poverty and participation poverty (Harris, 2004). The World Bank defines poverty more broadly than the income definition and includes powerlessness, noiselessness, vulnerability and fear, and sets a minimum expenditure threshold that categorizes those who live on less than US$ 1 a day as extremely poor. The European Commission looks at poverty as lack of income and financial resources, deprivation of basic capabilities and lack of access to education, health, natural resources and infrastructure (Mehta and Kalra, 2006).

Several studies have tried to explain the causes of poverty and a common denominator in all these is that poverty deprives and dehumanizes those affected. According to Marker *et al.,*(2002), the poor are not just deprived of basic resources. "They lack access to vital information about their lives and livelihoods: about market prices of the goods produced, about health, about the structure and services of public institutions, and about their rights. They lack political visibility, voice in the institutions and power relations that shape their lives. They lack access to knowledge, education and skills development that could improve their livelihoods. They often lack access to markets and institutions, both governmental and societal that could provide them with needed resources and services. They lack access to, and information about, income earning opportunities."

Gender Dimension of Poverty and Its Patterns in Kenya

In Kenya, the poor constitutes more than half of the population, *i.e.* at least one in every two Kenyans is poor (Anyang' Nyong'o, 2004). In monetary terms, absolute poverty in Kenya is pegged at KShs. 1,239 per person per month in the rural areas and KShs. 2,648 per person per month for urban areas of the country. According to the World Bank (2006), women in Kenya are poorer than men, with 54 per cent of rural and 63 per cent of urban women and girls living below the poverty line. Moreover, women are vulnerable to adverse shocks than men. This is due to the limited ability to own land and property, which negatively affects their ability to participate in economic growth. UNESCO (2006) commented on the issue of women and education, and noted that providing women with education is the only way of empowering them out of poverty. According to the United Nations (2006), if you educate a man, you educate one person, but if you educate a woman, you educate the entire family.

According to the Kenya Integrated Household Budget Survey (KIHBS), poverty manifests itself in various forms, with the prevalence and intensity of poverty among female-headed households higher than among those headed by males (KNBS, 2007). Of the poor, 75 per cent live in rural areas, while most of the urban poor live in slum and peri-urban settlements. The indicators of poverty have increased with the years, particularly in the education and health sectors. Households with a monthly income of Ksh. 23 671 and below are regarded as low-income households (KNBS, 2008). Some of the characteristics of the poor in Kenya include having large families, being engaged in subsistence farming, and lack of regular source of income. The GoK (2001) through the Poverty Reduction Strategic Paper 2001-2004 identifies several causes of poverty in Kenya as lack of basic socio-economic services like health and education coupled with inadequate credit facilities and gender imbalance.

Although all poverty eradication interventions need to be responsive to the specific needs and concerns for women and men in Kenya, unfortunately, lessons learnt from the trend of poverty situation in Kenya demonstrate that many of the Government's policies, objectives and strategies have either been gender blind or gender neutral. Policy makers have assumed that economic planning, programming and implementation impact equally on both women and men. They assume that targeting "people" is enough, yet, as Kimani (2006) puts that inclusion of women in poverty reduction strategies is the key element in achieving all the Millennium Development Goals (MDGs). Moreover, the achievement of Kenya Vision 2030 is pegged on the commitment as to identify and address the existing gender disparities in social, economic and political development.

Definition of Information and Communication Technology (ICT)

Greenbert (2005) offers elaborate definition of ICT and categorises it based on how long they have been in common use, and to some extent the technology used for the transmission and storage of information. New ICT includes computers, satellites, wireless one-on-one communications (including mobile phones), the Internet, e-mail and multimedia. The concepts behind these technologies are not particularly new, but the common and inexpensive use of them is what makes them new. Most of these, and virtually all new versions of them, are based on digital communications. On the other hand, old ICT includes radio, television, land-line telephones and telegraph. Old ICT has been in reasonably common use throughout most of parts of the world for many decades. Traditionally, these technologies have used analog transmission techniques, although nowadays they too are migrating to the less expensive digital format.

It is worth noting through Greenbert (2005) cautions, that the labels old and new are words that generally apply to the developed world. In many parts of the developing world, and in particular areas where literacy rates are low, ICT is effectively new. Although old technologies have their strength, this paper focuses mainly on new technologies, in particular such as: computers, internet, mobile phone and selected components of old technology. What stands out is that the mere availability of new technology does not translate to its use, neither does new technology automatically alleviate poverty. Several factors play together and around this scenario.

ICT in Kenya

Universal access to ICT has been identified by the government of Kenya as a major objective of the Vision 2030. It is expected that greater access to ICT will contribute to economic growth by reducing transaction costs and increasing businesses efficiency, especially in the case of small service firms in rural areas. In addition, ICT should contribute to the improvement of educational standards and access to information as well as accountability of government official Vision 2030. Part of the Vision is ICT based on its potential to increase productivity and raise the competitiveness of local businesses in a knowledge-based economy. From distribution and general standpoint, the government plan also considers that greater access to ICT can equalize opportunities and thus help to reduce income inequality. This can be done by supplementing and/or substituting certain forms of job training.

The Kenyan government together with external stakeholders and private contractors increase the ICT investment in order to reach the entire population regardless of demographic factors. Hallberg *et al.* (2011) conducted extensive research in the field of ICT to investigate the impacts of Digital Village Project (DVP) in order to understand their impact in rural areas. DVP is popularly referred as Pasha Centres. Pasha means "to inform", where the centers are located in rural and resource-poor environments. In the Kenyan context, digital villages are what normally other countries, *e.g.* in Sri Lanka and India, refer as telecentres (Hansson *et al.,* 2010). A telecentre in Kenya however, normally refers to what Jensen and Esterhuysen (2001) define as micro and mini telecentres. Therefore, a digital village in Kenya has a similar role as a telecentre in many other countries, *i.e.* to provide service through Internet and telecommunication media. In addition, digital village provides certain training, education, and e-government services.

The study revealed that although certain challenges exist in the project, women's access to ICT is higher in areas where the DVP is active. This means the success of such projects can be achieved if supported and encouraged to penetrate in all rural areas. It is also beneficial for women to embrace ICT when they are included in the specific targeted measures.

Development Patterns of ICT Sector in Kenya

According to the Report on ICT Access Gap (2011) by the Communications Commission of Kenya (CCK), one of the key factors in the development and expansion of ICT was the liberalization of the market that has started since 1999. The process has started with the splintering of the national public operator, the Kenya Post and

Telecommunications Corporation, into three different units: The Postal Corporation of Kenya, Telkom Kenya Ltd. (later privatized) and the Communications Commission of Kenya (CCK), which took on the regulator roles. This shift involved an important reorganization of the public ICT policy, from government monopoly to competition and the use of Private-Public Associations (PPAs) to increase the use of ICT. In 2004, the monopoly of Telkom Kenya Ltd. came to an end with an entrance of new private mobile operators, although it remained the sole fixed network operator in the country.

The Communication Commission of Kenya Report (2011) noted that in the case of internet, market liberalization was not as much of a factor as the development of appropriate infrastructure. While internet access in the country dates back from early 1990s, it was not until 1998 when the Kenya Posts and Telecommunications Corporation launched an internet access backbone, increasing the rate at which Internet Service Providers (ISPs) entered the Kenyan market. When the market was liberalized in 1999, however, all the informal ISPs as well as the newly privatized Telkom Kenya were granted licenses.

Access and Use of ICT in Kenya

The access to ICT can be defined as possession or availability (*e.g.* through home, office, school or public location) of ICT equipment, ability to pay for ICT products and services, and the skills to use ICT effectively. Survey results by KNBS (2011) point out a wide gap between the access of newer technologies such as the internet and that for traditional technologies such as telephony and radio.

Nowadays computers and other ICT gadgets have become part of the households in Kenya and are no longer perceived as a luxury. Households without the ICT facilities try to find the ICT services from nearby cyber cafés and friends' houses. In general, household ownership of personal computers, mobile phones and internet access are private and depend on the household incomes. Convergence of technologies in the ICT sector has also made easier access to ICT services such as the internet. Access, usage and ownership of ICT are the key factors in linking communities, facilitating businesses and empowering communities socially and for commercial purposes. It is therefore, essential that every effort should be made to bridge the digital divide between those who have information and those who need the information. Enhancement of access to information and communication services in rural, remote and underserved areas is the key element to accelerate the overall development as proposed by Duncombe and Boateng (2009).

The use of internet and mobile phones has witnessed a rapid increase, introducing radical changes to Kenya's' households. The growth of Internet and mobile phones usage in Kenya has been exponentially growing since 2007. Internet and mobile phones have become the basic means of communication for most Kenyans regardless of their economic status and geographical location. These technologies have increasingly become affordable to the lower socio-economic class of the population and is being used as a mechanism for greater participation of these groups in the development process. (Margaret N and T. Waema, 2011).

Access to voice services (mobile phones) is relatively high in Kenya. According to 2009 Census with more than 50 per cent of the population of Kenya had used mobiles phones. The CCK July-September 2012, a quarterly report, indicated a total of 30.4 million subscriptions in the mobile telephony market segment up from 29.7 million posted in the previous quarter, representing 2.5 per cent during the period. Similarly, the quarter recorded growth in mobile penetration to reach 77.2 per cent of the population which stood at 40 million according to the census. The 2012 CCK report thus shows a tremendous increase from the statistics of June 2009 that stood at 17.4 million mobile phone subscribers with an estimated population 38.3 million. In 2006 the number of subscribers was only 10.6 million.

Compared to voice services, access to data services (internet) is relatively low in Kenya. Following the results of 2009 Census, only 5 per cent of the Kenyan population had used internet during the previous month. This translated to about 2 million users. By 30th September 2012, there were 8.5 million internet subscriptions. The Census revealed that the popular places to get internet access in Kenya are cybercafés, work places, private houses and mobile phones. Cybercafés are mainly used in more developed counties like Nairobi and Mombasa, while in the less developed areas the popular internet access is by using mobile phones and telecommunication centres. Further, only 3.6 per cent of households have access to a computer. The CCK report further noted that mobile phones appear to be a more affordable alternative to the internet access due to their lower cost of equipment and lower operational costs (lower electricity consumption and do not require administrative personnel or maintenance).

According to the KNBS report (June, 2011), a higher proportion of males (7.5 per cent) compared to females (5.2 per cent) used internet. The report shows that frequency of use increased with level of education with proportions of population rising from 0.8 per cent for pre-primary to 59.5 per cent for those with higher education. This clearly explains that the low numbers of women internet users is because of less access to education when compared to their male counterparts. The access in rural areas remained low while the access is higher in urban population. The report stated that in urban areas one in every five persons used a computer within previous 12 months, while in rural population, only one in every twenty five persons used computers within the same period.

Although the internet use is not as rapid as the mobile phone use, there has been a remarkable upward trend over the years. According to the CCK (2009), the number of internet users by the end of 2008 stood at 3.65 million compared to September 2012 where it rose to 8.5 million. The number of internet subscribers recorded a growth of 34.5 per cent from a figure of 2.77 million recorded at the end of 2006.

For the development and maintenance of ICT, it is important to have other infrastructure like electricity, water, road, rail, and air transport systems in place. Electricity service in particular is viewed as a key driver for ICT and unfortunately less spread in rural Kenya especially in households headed by women. KNBS (2011) reported that only 1 in every 3 rural homes was connected to electricity as compared to 2 in 3 urban homes.

Huyer and Hafkin (2007) observed that a range of socio-economic and political factors affect and frame the gender divide. They include social and cultural barriers to technology use, education and skills level, employment and income trends, privacy and security, and available mode of access. These results support the findings that men and women differ in their usage of the new technologies, since skills are a prerequisite to usage. Skills are obtained mainly through formal education, although they are increasingly being acquired through other means as well.

According to the studies conducted by Zainudeen (2008) on mobile phone usage in Asia, there was a gender divide in access to ICT in Pakistan and India, but less of a divide in Sri Lanka, and none in the Philippines and Thailand, where women were empowered and mobile phones were pervasive. A conclusion drawn is that gender differences in usage are due to inequalities in the socio-economic domains, including education.

Benefits of ICT to Women

Gurumurthy (2006) indicated that there have been gains for females in the usage of ICT in many sectors, such as e-commerce, e-governance, health and information-sharing via the internet. However, she argues that the gains do not always give result in equitable gender relations. The KIHBS showed that the prevalence and intensity of poverty among female-headed households was higher than among those headed by males (KNBS, 2007). Gender per se may not influence the usage of ICT, but other factors related to gender such as inequalities in education or income levels, do affect the proportions significantly.

Terry and Gomez (2010) provide a qualitative study of the benefits of ICT and barriers facing women to gain ICT hubs fully. Their data classify the benefits as individual and collective benefits. Individually, women gain empowerment, increased self esteem, reduced isolation, access to markets and health information. Benefits accrued collectively include economic growth, improved health, capacity building and cultural transformation. The study also singled out location, infrastructure, connectivity, time constraints, money limitation, and lack of relevant context, low education, illiteracy and social norms as critical barriers to full realization of benefits.

It is important to note that internet and mobile phones users in Kenya are not necessarily the same group of people, with the difference attributed to the fact that mobile phones and the internet do not necessarily fulfill similar needs (Jagun *et al.*, 2007), *e.g.* offer reliable and cost-effective tools for serving households information needs (such as: conducting business transactions, communication and checking prices of products), households make choices on which technology to use based on the benefit and value derived from the usage. This is demonstrated through the rapid spread of the M-Pesa (mobile phone money transfer service) client base. According to Safaricom statistics, the M-Pesa service introduced in 2007 had 17 million Mpesa accounts with 2 million daily transactions as of April 2012. With the new mobile operators offering more value added services, the demand is expected to rise (Ndung'u and Waema, 2011). If every woman owned a mobile phone, then that would be a starting point that they can be enlightened on how to make use of other services besides M-pesa. However, the statistics reveal that ownership of mobile phones is skewed in favor of men in the country.

Challenges to Women's Access to ICT

A research study conducted by Odini.et al. (2012) on the topic of 'Empowering Rural Women in Kenya to Alleviate Poverty through Provision of Information: The Case of Vihiga District in Western Province', lists following hindrances to women's access to information:

1. Women were simply ignorant of the existence of suitable information
2. High level of illiteracy among women
3. Language barrier
4. Majority of women were so busy with their household chores that they simply lack time to seek required information.
5. Majority of women were so poor that they could not afford the cost of travel to get the required information
6. Infrastructure was unreliable and unsuitable with poor information services.
7. Older women found it difficult as a cultural practice to seek information from younger qualified experts.

The study by Odini *et al.* (2012) concluded that lack of access of information curtailed women participation in socio-economic development and rendered them ineffective in the fight against poverty.

If ICT has to be beneficial to reduce poverty, ICT need to be used to address real problems in sustainable way. This means that strategies should be put to focus on poverty reduction and not on ICT development. The starting point thus should be identification of measures to reduce poverty and then develop type of technology that would enhance success. When a new technology becomes available, one should figure out how it can be used to attack poverty. This approach can lead to innovative and effective use of new technologies. Ideally, ICT must have a positive impact on one or more aspects of poverty, affordability and also be accessible. Despite various pitfalls associated with the deploying of ICT projects, evidence grows that the use of ICT can be a critical and required component of addressing some facets of poverty. As reported by SIDA (2011), it is quite clear that ICT themselves will not eradicate poverty, but it is equally clear that many aspects of poverty will not be eradicated without a well thought-out use of ICT.

Results and Discussion

From the available literature on women in ICT and poverty, it is noted that Kenya is in the right track in empowering women through ICT. Several strategies have been put to bring women in the limelight thus bring about gender equity, political participation by women and eventually social economic empowerment. For example, the Constitution of Kenya promulgated in 2010 recognizes the role of women in economic development and protects their rights which initially suffered serious threats. According to the Constitution, women have the right to equal treatment including equal opportunities in political, economic, cultural and social spheres.

Parties to a marriage are entitled to equal rights at the time of the marriage, during the marriage and at the dissolution of the marriage which is a huge boost for women who previously lost everything in the event that the marriage did not work. Moreover the affirmative legislation which restrict single gender from occupying public offices means that women now have an equal opportunity. Women are in a positions to influence formulation of gender sensitive policies in the country, including policies related to ICT use and applications.

The government of Kenya recently has given a lot of thought to ICT use in the country. Of these, significant to note is the provision of laptops to all children joining class one in 2013. The current government formed by the Jubilee Coalition Party, has been referred to the "Digital Government" emphasizing its support to ICT. In this sense, knowledge will contribute positively to the development process in low-income households, through training and skill development, for maximum utilization of the technology.

The social pillar of Vision 2030 emphasizes the human resources of Kenya in order to improve the quality of life for all by targeting a cross-section of human and social welfare projects and programmes. It specifically mentions education and training, among several others. Further, the vision recognizes the overarching role of science, technology and innovation, and the need for enablers and macro foundations that include deploying world-class infrastructure facilities and services. There is a need to improve and expand rural infrastructure by focusing on shared access to facilities, with special focus on wireless technologies and required electrical power sources. Policy efforts should include the development and implementation of universal access funds to promote and support the deployment of wireless technologies and infrastructure in rural and remote areas, in coordination with electricity providers. The devolution of power found in the government structure can be used to ensure that rural areas too receive proper infrastructure which in turn offers a conduit for the establishment of ICT.

The digital village programme in Kenya need to be enhanced because of its potential to improve access and use of ICT. There is also a need for targeted and coordinated policies that aim to improve condition of life of women living in rural areas. The community-based access makes sense from an economic as well as social perspective. It provides affordable access without any expenses of ownership, as well as creates community setting for training and needed support service, including the needs of women and girls in rural communities.

It is important to develop and implement an education campaign focused on gender equality and women's rights within the context of ICT for development. It is critical to ensure that ICT does not become a reason for gender-based conflict and violence. There is a need for continuing public education around women's rights and gender equality. Adult literacy programmes in rural areas should be developed particularly for women to empower them stand against any form of violence. Promotion of programmes in local languages through radio, which are widely used, would be a proper approach to reach rural women.

There is a need to support collaborative initiatives to develop and disseminate local ICT content and its application in the areas of health, education, market information, agriculture, local administration and commerce, in a sustainable way so as to encourage faster adoption and productive use of ICT. To achieve this, public-private-NGO partnership, both for financing purposes as well as implementation and operational support, is critical. The partnership may include local or national businesses, including those providing technical support, rural cooperatives as well as other businesses with the ability to reach out to women.

Conclusions

There is a need for a national education and training framework to inform women of the available resources and the subsequent gains in tapping them, especially as the constitution and policies of Kenya have promoted women inclusion in the national development. Such empowerment would encourage them to actively participate in policy formulation, planning and programming in development process, and in return build support network of women for knowledge sharing. The framework would also guide in gender responsive mobilization and utilization of resources for effective engagement of women in all public and private sectors. It should also inform women on key strategies for sustainable economic growth and poverty reduction in all social, economic and even political areas.

The government should allocate resources for the expansion of informal sector and encourage the development of targeted women financial institutions in rural areas sensitive to the needs and abilities of women in respective areas. A framework needs to be developed on such institutions to remove possible interference and reduce unfair competition from established ones in urban and developed regions in the country. Besides, infrastructure should be developed to support the sector and enhance speedy, accurate and generally effective economic performance/gains in rural areas.

The local governments should mainstream gender in their counties and develop policies to bring women in the limelight. Such policies should be interactive enough to penetrate deep into the rural communities that have traditionally remained ignorant. Development in such areas would increase literacy in women and coupled with education and appropriate training, would bring about a paradigm shift from the traditional 'analogue' state to the current 'digital migration' fields by both the elite and the transformed illiterate.

References

1. Alkire, S. (2005), Why the capability approach?, *Journal of Human Development*, Vol. 6 No.1, pp.115-33.
2. Apoyo Consultoria, (2012), Study on ICT Access Gaps in Kenya: Final Report, July 2011.
3. (www.cck.go.ke) July-September 2012 quarterly report (accessed 2nd May 2013).
4. Communications Commission of Kenya, (April 2012) Second Quarter ICT Sector Statistics for 2011/2012, Nairobi, Kenya CCK Strategic Plan, 2008-2013. Communications Commission of Kenya (CCK). 2008.

5. David H., Mildred K., Ann K., and Loreen O. (2011), Case studies of Kenyan digital villages with a focus on women and girls. *Journal of Language, Technology and Entrepreneurship in Africa* Vol. 3 No. 1, 255 – 273 http://www.ajol.info/index.php/jolte/article/view/66724 (Accessed 5th May 22013).
6. Elishiba Njambi Kimani and Donald Kisilu (2010) Gender and poverty reduction: A Kenyan context. *Educational Research and Reviews* Vol. 5 (01), pp. 024-030.
7. Government of Kenya (GoK) (2007), Vision 2030: A Globally Competitive and Prosperous Kenya, Government of Kenya, Nairobi.
8. Gurumurthy, A. (2006), Promoting gender equality? Some development-related uses of ICTs by women, *Development in Practice*, Vol. 16 No.6, pp.611-6.
9. Greenbert A. (2005) ICTs for Poverty Alleviation: Basic Tool and Enabling Sector www.sida.se/publications (accessed on 29th April 2013).
10. Harris, R.W. (2004), Report on Information and Communication Technologies for Poverty Alleviation, UNDP-APDIP, Kuala Lumpur.
11. Huyer, S. and Hafkin, N. (2007), Engendering the Knowledge Society: Measuring Women's Participation, Orbicom, Montreal.
12. Kenya National Bureau of Statistics ICT Report (2011) in collaboration with Communications Commission of Kenya, http://www.cck.go.ke/resc/research.html (Accessed on 6th May 2013).
13. Kenya National Bureau of Statistics (2007), Basic Report on Well-being in Kenya – Based on Kenya Integrated Household Budget Survey, 2005/06, Kenya National Bureau of Statistics (KNBS), Nairobi, Kenya.
14. Kenya National Bureau of Statistics (2008), The 2008 Consumer Price Index (CPI),), Nairobi, Kenya.
15. Kimani E (2006), The Role of African Universities in the Achievement of Gender Equality and Women Empowerment, http://www.ku.ac.ke/schools/humanities/article/view (Accessed on 12 May 2008).
16. Marker P., K. McNamara and L. Wallace, (2002). Report on the significance of information and communication technologies for reducing poverty.
17. McNamara K.S. (2003), Information and Communications Technologies, Poverty and development: Learning from Experience, infoDev; http://www.infodev.org/files/833_file_Learning_From_Experience.pdf (downloaded 20th April 2013).
18. Margaret Nyambura Ndung'u, Timothy M. Waema, (2011), Development outcomes of internet and mobile phones use in Kenya: the households' perspectives, *Info*, Vol. 13 Iss: 3, pp.110 – 124.
19. Mehta, S., Kalra, M. (2006), Information and communication technologies: a bridge of social equity and sustainable development in India, *International Information and Library Review*, Vol. 38 No.3, pp.147-60.
20. Ndung'u, M.N., Waema, T.M. (2011), Development outcomes of internet and mobile phones use in Kenya: the households' perspectives, *Info*, Vol. 13 No.3, pp.110-124.

21. Odini Serah, Otike, Japhet and Kiplangat, Joseph: Empowering Rural women in Kenya to Alleviate Poverty Through Provision of Information: The Case of Vihiga District in Western Province Presented at SCECSAL Conference hosted by KLA on 4th-8th June 2012 venue LAICO REGENCY Hotel Nairobi, Kenya.
22. Robeyns, I. (2005), The capability approach: a theoretical survey, *Journal of Human Development*, Vol. 6 No.1, pp.94-114.
23. Terry, A. and Gomez, R. (2010). Gender and public access computing: An international perspective. *The Electronic Journal of Information Systems in Developing Countries*, 43(5), 1-17.
24. World Bank (2007), Gender and economic Growth in Kenya: unleashing the power of women, Washington, D. C.: The World Bank.
25. World Bank (2006), Information and Communications for Development. Global Trends and Policies. Washington, D.C.: World Bank.
26. Zainudeen, A. (2008), What do users at the bottom of the pyramid want?, in Samarajiva, R., Zainudeen, A. (Eds),ICT Infrastructure in Emerging Asia: Policy and Regulatory Roadblocks, IDRC/Sage, New Delhi/Ottawa, pp.39-59.

Chapter 8

ICT and Women's Empowerment: A Paradigm Shift Needed in Outlook

Archita Bhatta

Science and Environment Journalist,
Scidev.net, alertnet.org,
New Delhi, India
E-mail: architabhatta@gmail.com

ABSTRACT

ICT has proved to be a very effective tool in the empowerment of rural women because it bridges distances in giving them access to services, information and to the market. It also creates groups which can work together in several areas, like exchanging information, helping each other and also developing small businesses together.

However, such stories have remained buried in the celebration of the success of ICT and economic growth, which has overwhelmed the media ever since India's ICT boom. The utility of ICT in development in general, and women's empowerment in particular, has remained largely in *purdah*. This finds mention in a limited number of discussions and reports, especially of development agencies and non-governmental organizations, and is not reflected in the media coverage on ICT.

ICT is still largely viewed as an urban facility and empowering women through ICT is largely perceived as education in computer and related sciences to help educated women secure jobs in the ICT sector.

Media can be a crucial lever in changing such an outlook. This paper looks at the role that media can play in promoting ICT as a tool to empower women, and the steps that need to be taken to change the outlook.

It also shows how interest can be generated in the media on the importance of ICT in women's empowerment by drawing from examples of communication on issues such as climate change. The importance of communication tools in bringing this about has also been highlighted in the paper.

Introduction

Information today can be created, disseminated, stored, shared, value added to and managed with the use of a variety of technologies which are ever on the increase. All these, including the ones that are being added everyday encompass information and Communication Technology (ICT). They include telecommunications, television and radio broadcasting, computer hardware, software and services and electronic media, for example, the internet, electronic mail, geographical information systems (GIS), web-based GIS and many more.

ICT can play a major role in bridging distances to provide access to information for better freedom of choice, improved health care and also to better economic and social opportunities especially to the underprivileged people (Walshal, 2010). Rural women have less opportunities, especially the ones who are poor, have limited education and mobility, less freedom to interact with outsiders, no financial independence and no decision making power. Hence, they are one of the most underprivilegedsections of the society.ICT can play a major role in helping them to access information, healthcare, education, social and economic opportunities, and facilitate in helping them to take their own decisions.

ICT not only can help in accessing services, it can also help them to interact with a diverse range of people, reach out for help and also form groups, and learn from each other. Such groups can exchange informationand start small businesses together. ICT effectively holds a great potential in the empowerment of rural women.

Despite this potential, ICT development in India is characterized by a gender based digital divide. The 2004 report by the Cisco Learning Institute says that women comprise only 23 per cent of India's internet users. Low levels of access to ICT have resulted in a gender digital divide in India. The most significant reasons behind this are poverty, lack of computer literacy and language barriers which hamper access especially in developing countries (Gurumoorthy, 2004, Barbara J, 2007,Sue Lewis *et al.,* 2007and Huyer and Sikoska, 2003).

Success Stories Exist, but Remain Unmentioned in Media

Notwithstanding the hurdles, there are several examples of empowerment of rural women in India through ICT. Women in Gujarat have immensely benefited from the Dairy Information System Kiosk, which manages detailed inventory of milk cattle in the state, and also provides information about veterinary and otherservices relevant to the dairy sector (Jain, 2006).

The Networking Rural Women and Knowledge, a UNESCO project, has documented such cases. For example, the information on: income-generating activities, specific education projects, microfinance and health, have been shared between women in Baduria in North 24 Parganas district of West Bengal. As a result of this exchange of information, more and more young educated women have obtained access to and control of ICTs while less educated and older women have benefited from the availability of access to information.

In some places people have used ICT to market their village handmade handicrafts and other products internationally. These women said that knowledge of ICT has helped them to gain greater respect in their villages, to express their creativity and also to contribute to their family income.

However, it is not only just about computers and the internet, the mobile phone and the community radio have played a crucial role in reaching agricultural information to women. ICT helps them to know about the latest developments in their fields of interest, as well as to share their problem and solutions.

For example, the mobile phone has played a key role for women in Uttarakhand to share problems and solutions related to their food product-based business in the state, and also to improve their products (personal experience, 2008).

ICT has also become a powerful medium through which educated women from rural backgrounds can highlight problems faced by the local community. For example, a woman from Muslim weaver family in Bihar has been trained in ICT and she made a video on the plight of Muslim women weavers in her village. The award winning film has been able to highlight their problems through YouTube.

However, such case studies of women's empowerment through ICT are rarely highlighted by the national or international media.

Where in the Media?

A study of the coverage of ICT by the media in Malaysia analyzed 299 news from the significant newspapers in the country. It showed that most stories come in the form of hard news, economic, governmental and international issues (Mohd Yusof Hj. Abdullah *et al.,* 2009). There were 75 news on ICT and business, commerce and economy, 56 on ICT and development, management and services, 45 on ICT and education, 35 on ICT, science and technology. Human interest stories and those on policy, rural development and women's empowerment were negligible. The newspapers were actually more informative, but not educative or persuasive.

A study by the World Institute of Development Economics shows a similar overemphasis on ICT and economic growth in India. India's ICT growth story is mainly confined to export oriented growth. The government highlighted that how export of ICT products and ICT based human resources has brought about economic growth, and the media has been overwhelmed by it (Joseph K J, 2002,Shobha Arun *et al.,* 2004).

While government policies mainly aimed at promoting economic development through ICT, the media celebrated this export oriented growth, thus, ICT for development received limited attention. It is relegated to a limited development fund and hence there is scant reflection of what ICT can do for rural communities let alone rural women (Singall and Rogers, 2001; Heeks, 2002; Shobha Arun *et al.,* 2004).

As a result, policies have not been shaped to ensure affordable and effective ICT access for women (Nancy Hafkin, 2002, Sonia Jorge, 2002, Anita Gurumoorthy, 2004.Barbara J 2007,Sue Lewis *et al.,* 2007,Huyer and Sikoska, 2003).This is because the initial purpose behind information technology was not for development. It was driven by commercial interests (Anita Gurumoorthy, 2004).

Development came much later as a by-product. It was realized even later in the late 1990s that it can be an effective tool for women's empowerment (Annan, 2005).The action on this started only after it was realized that ICT for women's empowerment requires affordable and effective access (Annan, 2005).

The early 21st century saw the importance of promoting the use of ICT for women's empowerment. It gained in the dialogues of development agencies and international organizations like the United Nations. For example, one of the targets of the Millennium Development Goals (MDG) was to make available new technology benefits specially ICT (Huyer and Mitter, 2003).It was also said that mainstreaming of ICT would contribute to achievement of the MDG (Huyer and Mitter, 2003). The UN division for advancement of women mentioned that ICT with enabling environment can provide avenues for women's empowerment.

The meaning of empowerment also evolved with time and along with it the definition of affordable and effective access (Nath, 2001). For example, until early part of the 21st century, women's empowerment through ICT was understood as media, communications, and networking. In the mid 21st century years, it was seen as a tool of economic and social empowerment for underprivileged women; and now empowerment entails enhancing capacity to influence and participate in making decisions influencing their lives (Garba, 1999). Accordingly, the level of affordability and access needed to bring about such empowerment is changed (Nath, 2001).

Drastic changes in policies are needed to ensure this level of affordability and access. This is because the access is determined by several factors like: availability of electricity, internet access, frequency access, transportation facilities, feeling of security, location of points of usage and also understanding of the nature of women themselves (Hafkin, 2002, Nath, 2001,Jorge, 2002).

It requires a drastic change in the mindset of policy makers and interdepartmental coordination (Hafkin, 2002;Nath, 2001;Jorge, 2002),where in developing countries would take a lot of time unless it is driven from the top. Such that the policies requirements failed to catch up with these changes (Hafkin, 2002; Jain, undated).

This 'Limited Edition' Outlook Needs Media Attention

One of the reasons behind the lack of change in mindset was that the discussions about the need to promote ICT for women's empowerment were confined to discussions and reports of a few development agencies and non-governmental organizations (Huyer and Mitter, 2003).

Outside that circle, talks about it were a rarity, reflections in academic circles were a rarity, as were among ICT experts, and deliberations were mostly confined to grey literature (Jain, undated). With a lack of discussion among experts, the topic rarely found a place in the media (Malaysia, ICT and development, undated).As the media rarely touched upon the topic, awareness about it among the public failed to increase.

Media can be a crucial lever in changing this attitude. If the media is involved in creating awareness and interest in the issue, public demand for action increases. Policymakers are sensitized and more parliamentary debates and questions can be generated along with public debate on the topic. The questions in parliament from public have to be answered and hence action will be initiated. This will be reflected in public policy changes and implementation of these changes.

Conclusions

Media Needs Sensitization

In order to achieve the above actions, the media needs to be sensitized. The current attitude of the media towards ICT and women's empowerment needs to be assessed, and methods have to be found to change this attitude.

Positive stories on empowerment of rural women through ICT should be highlighted to the media by triggering media interest in such stories. NGOs working in this field should be proactive in highlighting these stories.

New tactics have to be evolved to increase media coverage of ICT and women's empowerment. For example lessons can be taken from climate change coverage by media. Studies show that media covers an issue in which national and international policy makers talk about it (Boykoff 2010, Boykoff 2007, and Boykoff and Boykoff, 2007).

Taking a clue from this, ways can be found to involve important national and international politicians to comment on women empowerment by ICT. Means have to be found to introduce the issue in their speeches on significant events like the women's day, technology day and so on.

Studies also show that climate change coverage increases significantly during CoP meetings (Boykoff 2010; Boykoff 2007 and Boykoff, and Boykoff, 2007). Taking this as an example, international dialogues and debates need to be organized on such issues and problems faced in the issue put to the forefront of the particular events.

Besides, occasions like the environment day and earth day have found climate change coverage increase several times. ICT and women's empowerment can also have such designated days for revisiting the issue each year and development agencies can organize communication strategies to popularize such issues around the event.

References

1. Geoff Walsham (2010). ICTs for the broader development of India: An analysis of the Literature; Judge Business School, *Electronic Journal on Information Systems in Developing Countries*, 41(4): 1-20.
2. Mohd Yus of Hj. Abdullah, Fuziah Kartini Hassan Basri, Mohd Safar Hasim and Mat PauziAbd Rahman (2009). The portrayal of ICT in the media: a

Malaysian Scenario, UniversitiKebangsaan Malaysia, *JurnalKomunikasi, Malaysia Journal of Communication* Vol 25: 1-12.

3. Joseph, K. J. (2002). Growth of ICT and ICT for development: Realitiesof the myths of the Indian experience, WIDER Discussion Papers//World Institute for Development Economics (UNU-WIDER), No. 2002/78, ISBN 9291902810.
4. Jain Suman (2006). ICTs and women's empowerment: Some case studies from India, Department of Economics at Lakshmibai College, Delhi University.
5. Anita Gurumurthy (2004). When Technology, Media and globalization Conspire: Old Threats, New Prospects;Panel on globalized media and ICT systems and structures and their interrelationship with fundamentalism and militarism; Isis International-Manila, World Social Forum in Mumbai, India.
6. Barbara J. Crump, Keri A. Logan and Andrea McIlroy (2007). Does Gender Still Matter? A Study of the Views of Women in the ICT Industry in New Zealand, *Gender, Work and Organization*. Vol. 14 No. 4 July.
7. Sue Lewis, Catherine Lang, Judy McKay (2007). An Inconvenient Truth: The Invisibility of Women in ICT; *Australasian Journal of Information Systems Volume 15 Number 1 December.*
8. Sophia Huyer and TatjanaSikoska (2003). Overcoming the Gender Digital Divide: Understanding ICTs and their potential for the empowerment of Women by, Instraw Research Paper Series no 1. April 2003.
9. Sonia Nunes Jorge (2002). The Economics of ICT: Challenges and Practical strategies of ICT use for Women's Economic Empowerment, United Nations Division for the Advancement of Women (DAW) Expert Group Meeting on "Information and communication technologies and their impact on and use as an instrument for the advancement and empowerment of women" Seoul, Republic of Korea, October 2002.
10. Nancy Hafkin (2002). Gender Issues in ICT Policy in Developing Countries: An Overview; United Nations Division for the Advancement of Women (DAW) Expert Group Meeting on "Information and communication technologiesand their impact on and use as an instrument for the advancement and empowerment of women" Seoul, Republic of Korea; October 2002.
11. P. KasseyGarba (1999). An endogenous empowerment strategy: A case-study of Nigerian women; Development in Practice, Volume 9, Issue 1-2, 1999.
12. Sophia Huyer and SwastiMitter (2003). ICTs, Globalisation and Poverty Reduction: Gender Dimensions of the Knowledge SocietyPartI. Poverty Reduction, Gender Equality and the Knowledge Society: Digital Exclusion or Digital Opportunity? (UNCSTD).
13. VikasNath, (2001). Empowerment and Governance through Information and Communication Technologies:Women's Perspective; *Intl. Inform. and Libr. Rev.*, 33, 317-339.

14. ShobhaArun, Richard Heeks and Sharon Morgan (2004). ICT Initiatives, Women and Work in Developing Countries: *Reinforcing or Changing Gender Inequalities in South India?,*Development Informatics, Working Paper Series, paper no. 20.

15. Singall, A. and Rogers, E. (2001). *India's Information Revolution: From Bullock Carts to Cyber Marts,* Sage Publications, New Delhi.

16. Heeks, R.B. (1996). *India's Software Industry,* Sage Publications, New Delhi.

17. Kofi Annan (2005). Gender Equality and Empowerment of women through ICT, Women 2000 and beyond; United Nations Division for the Advancement of Women, Department of Social and Economic Affairs.

18. Max Boykoff (2010). Indian media representations of climate change in a threatened journalistic ecosystem, Climatic Change 99:17–25.

19. Maxwell T Boykoff (2007). From convergence to contention: United States mass media representations of anthropogenic climate change science, *Trans Inst Br Geogr*NS 32 477–489.

20. Maxwell T. Boykoff, Jules M. Boykoff, (2007). Climate change and journalistic norms: A case-study of US mass-media coverage, *Geoforum* 38 (2007) 1190–1204.

21. http://www.youtube.com/watch?v=Xq87aV1oO-o.

Chapter 9

Harnessing the Power of New ICTs for Rural Women in India: NGO Roles

Jane Schukoske

CEO,
S.M. Sehgal Foundation, India
E-mail: j.schukoske@smsfoundation.org

ABSTRACT

Information and Communications Technology (ICT), important for everyone, holds special potential for rural women. Rural women often have less education, less access to information in their daily lives, and greater isolation in domestic settings. In some communities, women and girls may lack freedom to attend meetings, find answers to their questions, and speak with outsiders, especially male ones. Rural women are using ICTs to surmount disadvantages: to learn, enjoy better health, prosper in livelihoods, access legal help and take care of government and private transactions.

Non-governmental organisations (NGOs) working in rural communities in India routinely support traditional community media such as local language newsletters, posters, pamphlets, street theater (*nuked natak* in Hindi), wall paintings and community meetings. Rural women can relate to ICT materials in local language and with pictures that convey messages that are relevant to their daily lives.

"New ICTs" are accelerating the ability to reach out to villagers in rural areas and to provide additional platforms for interaction. New ICTs, which may be used individually or in convergence with each other, include community radio, Internet radio, mobile phones, computers/laptops/tablets and other handheld devices, information kiosks, tele-centres, video cameras, and the Internet (Nath, 2001). There is an abundance of pilot projects using ICT to engage rural communities. Challenges to engaging rural women through ICT are many – from infrastructure inadequacies to patriarchal social structures and honing in on relevant and engaging content.

This paper begins with discussion of the roles of NGOs in India in spreading new ICTs to empower rural women having little formal education. We then examine one Indian NGO's experience with introduction of ICTs in development work, particularly with community radio, and three other Indian examples of the many ICTs used to engage rural women – mobile phone applications, multi-purpose village kiosks, and interactive voice response systems. The paper concludes with identification of challenges and recommendations on next steps for NGOs in India to accelerate use of these technologies for empowerment of rural women.

Keywords: *Non-government organisation, Women's empowerment, Community radio, Information and communication technology, Rural development, India.*

Introduction

Roles of Community-Based NGOs

NGOs working at the grassroots encourage community participation so that people are better able to address their own development needs. Most rural development NGOs focus on particular areas, such as education, governance, health, livelihoods, water and sanitation, and women's empowerment. Whatever the development focus of an NGO working in communities, each NGO has a role in ICT at the least to promote their main educational and cultural messages.

Development practitioners widely accept the fact that women's involvement is essential for social and behavioral change in communities due to their focus on the wellbeing of their families. Women living in rural India often have little education and access to ideas about solutions for their family and village problems. In some communities, women have little public presence. It is essential for NGOs to focus on the special situations of rural women in the areas they serve and to reflect on what more the NGO might do to reach women and facilitate greater participation by them. Reaching out and engaging women through ICT can contribute to this effort.

NGOs serve as a bridge for communities to information about government programmes and to initiatives that other communities are taking. NGOs often help individuals and communities build skills as well as knowledge. They help communities take stock of the their situation relative to other villages, and to plan for improvement. In India, they often build capacities of local government institutions such as *panchayats* (village councils) and village committees on education and health. Some NGOs support formation of women's Self Help Groups (SHGs) and federations of SHGs. NGOs often identify resources to fund/bring about change – government and donor programmes, collaborations to generate income, and community contribution to development projects. NGOs also teach citizens skills for monitoring of government programmes.

In 2013, NGOs in India are proactively piloting new ICTs in rural communities while at the same time advocating for and staying alert to emerging government ICT efforts. Under the Universal Service Obligation, the government of India is obliged to "provide access to telecommunications services to people in rural and remote parts of the country at reasonable and affordable prices (Gulati, 2010)." The Universal Service Obligation fund has taken gender-specific initiatives in compliance with the requirement of gender responsive budgeting (Gulati, 2010). In terms of access to government services through ICT, the National Portal of India serves as a repository of materials on e-governance at a level for government officials and citizens seeking government websites (Voluntary Association for People Service, 2011). At the central level, Ministry of Health and Family Welfare started the Mother and Child Tracking System started in 2010 and, at the state level, Uttar Pradesh Department of Food and Civil Supplies has piloted a system for announcing food grain delivery as part of the Transparent Pargeted Public Distribution System (Digital Empowerment Foundation *et al.*, 2012).

In light of the widespread poverty and the need to accelerate development efforts in India, NGOs have the opportunity and responsibility to bring ICT to communities in which they work. Grassroots intermediaries such as NGOs and community-level health and education workers are key to success in poverty reduction through ICTs (Cecchini *et al.*, 2003). It is incumbent on NGOs to model, or collaborate with others who can model, useful and feasible technology to serve and engage rural communities.

With many pilot ICT projects underway in India, it is important that NGOs collaborate to share their ICT strategies, experiences and learnings. The Digital Empowerment Foundation has established the Digital Knowledge Centre website, http://digitalknowledgecentre.in/, as a platform to promote sharing of ICT pilots and strategies in a broad range of development fields. Mobile phone technology is a main focus of NGOs given the widespread use of mobile phones even in rural areas of India; its widespread use provides the potential for scalability of pilots showing good impact (Digital Empowerment Foundation *et al.*, 2013).

Materials and Methods

This paper begins with the first-hand NGO experience of S M Sehgal Foundation in Gurgaon, Haryana, India, in piloting ICTs in rural communities as part of its core development work on water and sanitation, governance, capacity building and agricultural income enhancement. It then discusses literature regarding use of new ICTs for rural women's empowerment and community development in India. The paper concludes with identification of challenges and recommendations on next steps for NGOs in India to accelerate use of these technologies.

Case Study of SM Sehgal Foundation

S M Sehgal Foundation ("Sehgal Foundation"), a trust registered in India in 1999, dedicates its efforts to further the wellbeing of rural communities in India. With community participation as its approach to insure relevance and sustainability, Sehgal Foundation develops, tests and spreads models on water management, agricultural income enhancement and good governance. Sehgal Foundation has worked in rural development for over a decade primarily in villages in Mewat district, Haryana, where human development indicators are low, and it implements projects in neighboring Alwar district, Rajasthan.

Sehgal Foundation's Communications department publishes Hindi language newsletters for the communities in which it works and its programme staff work with communities on wall paintings to publicly share information in the villages. Community radio station *Alfaz-e-Mewat*, FM 107.8, broadcasts from Sehgal Foundation's community center in Ghaghas, Mewat District, Haryana. A link to live streaming of radio broadcasts is on Sehgal Foundation's home page, www.smsfoundation.org.

Sehgal Foundation observes that many rural residents, many more men than women, in Mewat use mobile phones and very few use computers or Internet services. Below are descriptions of three of Sehgal Foundation's efforts to integrate new ICT into its work: computer-assisted learning, mobile phone supported advice for farmers, and community radio.

Figure 9.1: Mamta: An Alfaz-e-Mewat Radio Jockey from the Local Community, Speaks on Air.

Laptop Assisted Adult Literacy Project

Sehgal Foundation piloted TARA Akshar in Firozepur Jhirka and Nagina blocks of Mewat district from March to December 2007. TARA Akshar, a project of the Development Alternatives Group, at that time taught Hindi literacy to children and adults in 30 days. Students learned with a teacher using laptops with customized Hindi software, based on memory techniques, flash cards and reading materials. The participants were in the age group of 7 to 40 years. Participation in the project was high; there were few drop-outs. The project achieved the literacy goals for the rural girls and women who participated. After the pilot, Sehgal Foundation discontinued the project in Mewat district due to cost. However this decision may be reconsidered in a future planning cycle. In 2013, the TARA Akshar programme is now 49 days long and operates in parts of Uttar Pradesh and Rajasthan.

Mobile Phone Supported Advice for Farmers

Sehgal Foundation and One World South Asia piloted the LifeLines Agriculture project in 2010. LifeLines provided expert advice to farmers through mobile telephones on agricultural practices, seeds, market situation, prices and more. The project resulted in improved soil health which led to increased productivity, increased earnings and savings. There were other favorable changes such as better nutrition, disease control, improvement in education and skill enhancement (Mehta, 2010). There were several learnings from the pilot. Agricultural information is region specific and it is very crucial that the farmer receives information that is relevant to his/her geographic location, crops and weather conditions. Timeliness in response to farmer's questions was critical to success of the programme. Delays in response discouraged participation in the project. The pilot project in Mewat ended with the funding. Sehgal Foundation is drawing on the learning from this project to refine a model for a pilot with women farmers.

Community Radio

Rural community radio stations serve the dual function of sharing information with listeners and providing the listener community with a means for voicing concerns and opinions. Through a toll-free number or a dedicated phone line, radio listeners can call in to respond to polls or to listen to specific radio programmes. Many people in rural areas access community radio through their cell phones.

Sehgal Foundation's community radio station *Alfaz-e-Mewat* (Voice of Mewat), FM 107.8, is one of 144 community radio stations operating in India (Ministry of Information and Broadcasting, 2013). *Alfaz-e-Mewat* broadcasts for almost 11 hours per day during the morning and evening hours that are the most convenient times for listeners. Caller records indicate that listeners hail from 183 villages in Mewat district and 10 villages in Alwar district, Rajasthan. The radio programming is designed to reach out to listener groups in the community, particularly women, children, youth and farmers. Programming includes shows on agriculture, health, good governance, education, water, entertainment for children and live shows for the community at large. The community radio staff are local people who were trained for their roles at the station, including equipment operation and maintenance, reporting, broadcasting, and editing.

The history of setting up *Alfaz-E-Mewa*t illustrates the complexity of the work of an NGO in supporting such an ICT project. Sehgal Foundation applied for a community radio license in 2009 and the license was granted in 2011. The station began broadcasting in February 2012. Funding was provided by the state Ministry of Agriculture for station construction and for three years of operation, after which time the station is expected to be self-sufficient through advertising, sponsorships and community support.

Sehgal Foundation also recruited another partner, Sesame Workshop India Trust, which provided the broadcasting software (Grameen Radio InterNetworking System, GRINS), training and 90 episodes of a children programme called *Galli Galli Sim Sim* (an Indian adaptation of Sesame Street in the US). After broadcasting for more than the three-month minimum requirement, Sehgal Foundation has been empanelled with Directorate of Advertising and Visual Publicity to run paid public service announcements of government departments. This revenue contributes to the sustainability of the station, which will be essential once the initial grant support ends.

Community radio station operation energizes both Sehgal Foundation's staff and the community. The presence of the radio as a community institution has built hope. There are opportunities for learning, reviving and enjoying local culture, and discussing important local development issues – agriculture, water management, good governance including access to entitlements and services such as education and health care.

Call-in radio shows allow interaction with experts and other guests. The radio station employs listener call-in campaigns to obtain feedback on programming and to elicit ideas and opinions from listeners.

Its community radio experience has led Sehgal Foundation to provide a platform for many organizations working on community media to share experience and learn from the expertise of others. Sehgal Foundation partnered with UNESCO in 2012 to convene an interstate conference in Gurgaon entitled "Rural Voices: Unheard to Empowered" that featured panel discussions and a field trip to the *Alfaz-e-Mewat* station. In view of the wide interest in such a platform for sharing ICT information, Sehgal Foundation and UNESCO held a second annual conference on mainstream and community media in 2013.

The Gender Policy passed by the trustees of the Sehgal Foundation in 2012 has prompted closer attention to outreach to women through ICT. The policy calls for assessment of all of the Foundation's programming as well as other aspects of operations to better accelerate women's empowerment. For example, discussion of this policy sparked the idea for a radio call-in campaign before International Women's Day to elicit nominations from listeners of unsung women achievers in their villages to be publicly recognized.

As of July 2013, some ICT based projects that are in the pipeline at Sehgal Foundation include the promotion (via voice and text messages) of a bio-sand filter for purifying water; use of a flash movie clip played on a tablet computer to educate villagers on hygiene and sanitation; support for farmers with text messages on weather forecasts, and text messages on statutory entitlement updates for citizens in a training programme on governance. Apart from these pilot projects in the community, Sehgal Foundation has planned an online-assisted training programme for the NGO staff to improve their English language skills. On the face of it, the content of these projects is gender neutral, though we plan the messages to be heard by women, who are more likely than men to take action in the interest of their family and community health and wellbeing.

Uses of New ICTs in Empowering Rural Women: Examples

New ICTs offer the potential to reach rural women who are isolated within their family and village structures. New ICTs not only perform traditional functions of sharing information and providing means of communication, but also add speed of outreach, the opportunity for interaction across distances, and the opportunity to reach much larger audiences. ICT software provides the capacity to sort information, target messages to particular audiences, and track the messages and responses. Pictorial and voice messages are especially important for reaching women in view of the high rate of illiteracy among rural Indian women.

In addition to the community radio technology described in the Sehgal Foundation case study, three other categories of ICT that hold promise for rural women's empowerment are discussed below: mobile phone technology, village kiosks or centres, and interactive voice response systems. These technologies are noteworthy for specific reasons: mobile phones are widespread and allow interactivity; village kiosks provide local access without the need of an individual mobile phone and also provide a place for social interaction that tends to reinforce use of the technology and learning; and interactive voice response systems tap the interests of the rural women users, ensuring that the content is relevant and the communication level appropriate.

Though these are examples of freestanding technologies, there is increasing convergence in their use. Community radio may or may not have an interactive voice response system component, for example. Interactive voice response systems may be accessed by mobile phones or landlines.

Mobile Phone Technology

Women's ownership of mobile phones in India lags behind men's ownership but is rapidly growing. From 2009 to 2011, women's ownership of mobile phones grew by 40 per cent (Digital Empowerment Foundation and UNICEF, 2013). Fewer Indian rural women than urban women own a mobile phone. The reasons women give for not owning a mobile cited in a 2010 study by Vital Waves Consulting include "the cost of handsets and service, a lack of need for a mobile phone and fear of being able to master the technology. Cultural issues, such as the traditional roles of men and women, are also a factor in women's mobile phone ownership and can delay or even prevent a woman's acquisition of a mobile phone" (Vital Waves Consulting, 2010).

Experts in India have identified six ways in which communities, NGOs, government offices and others are using mobile phones in rural development work:

- ☆ Information dissemination
- ☆ Interpersonal communication
- ☆ Progress tracking, monitoring
- ☆ Training of front line workers
- ☆ Advocacy and outreach
- ☆ Community mobilization (Digital Empowerment Foundation *et al.*, 2013)

A few examples of mobile phone projects of particular relevance to rural women are:

Helplines that provide information and interaction: This is a traditional use of telephone for support to the public. Mobile phones make it possible to place calls from locations away from landlines, either in or outside homes or at community sites, which may increase use of the helplines and thoroughness and accuracy in reported information. Two examples are for a women's helpline addressing security issues for women through a collaboration between an NGO and a police station (Vodafone India Foundation *et al.*, 2011), and a helpline of the Haryana Department of Education to accept reports of complaints about implementation of the Right to Education (Department of School Education, Haryana, 2013).

Short messaging service (SMS) audio or pictorial technology: This is a way to broadcast locally relevant information to targeted audiences regardless of illiteracy. Such SMS are used in many kinds of projects, including for:

(a) Accessing information such as weather or market information for women farmers (Vodafone India Foundation and Digital Empowerment Foundation, 2011),

(b) Sending health messages to pregnant women and new mothers about maternal and child health (Vodafone India Foundation and Digital Empowerment Foundation, 2011),

(c) Transmitting information about local water availability (NextDrop, 2013)

(d) Training women in goat rearing and negotiating skills as business women (Vidiyal, 2012; Spaven, 2011), and

(e) Providing updates on relevant to their business opportunities (Digital Empowerment Foundation *et al.*, 2013).

In some communities, women may perceive no need for a mobile phone in their daily routine. NGOs can build awareness of the potential benefit to women by spreading concrete local examples of the helpfulness of timely information.

Village-Based Centers or "Kiosks"

Where can rural women access technology besides through mobile phones?

In the developed world, there is often Internet access through public libraries; a consortium of organisations is spreading this approach in developing and transitioning countries (Beyond Access, 2012). In the absence of such public library facilities in rural India, some NGOs, local governments, NGOs and entrepreneurs have helped establish village-based centers or "kiosks". These "kiosks" may be located in existing institutions – such as health centers, schools and community centers (United Nations, 2005). With computer access and skills, rural women can use e-governance tools for information, requests and transactions (Government of India, 2008). Some of rural women with sufficient education can be trained for self-employment in the IT sector, such as data management and distribution transaction processing (Sulaiman *et al.*, 2011). Such rural internet kiosks, sometimes community-owned, can save villagers' time and money. Gyandoot, for example, is an intranet network supported by *Panchayats* in Dhar, Madhya Pradesh (Gyandoot, 2012). Some "kiosks" in India are also being utilized to improve the quality of education, get better job opportunities with higher salaries, and enhance literacy and awareness (Voluntary Association for People Studies, 2011).

For rural women to effectively utilize ICT in their daily activities, they require training and a continued support structure at least in the initial stages. Training programmes should closely monitor participants' learning, level of self-confidence, skills and knowledge and overall give moral support. Women will feel empowered only if they are able to clearly see the benefits of using ICT and improve the quality of their lives. For women with secondary level of education, training in ICT has contributed to expanding income-generating opportunities by providing new forms of employment opportunities. For example, the Credit Society of Medchal (Andhra Pradesh, India) has been training its women members on digital literacy and encouraging them to start computer training centres as business propositions (Khalafzai *et al.*, 2011).

Some researchers have found that community radio and village knowledge centres, which have locally relevant content and reinforce community-based discussions, have the greatest potential to reach women (Sulaiman *et al.*, 2011). On the other hand, researchers have pointed out that unless the underlying government services, often performed manually, are improved, that the technology does not bring significant change (Indian Institute of Management, Ahmedabad, 2002).

Interactive Voice Response System

ICT content should be relevant to rural women, expressed in local parlance and provided in a timely fashion. One approach to insure relevance and local language is to start the flow of information from the community, engaging community members in expressing their thoughts, questions and concerns and in responding to the matters raised by others in their community (Cecchini, 2003; Gram Vaani, 2013). The social technology organization Gram Vaani (Voice of the People) created Jharkhand Mobile Vaani to provide such a platform, built on an intelligent interactive voice response system that allows people to call into a number and leave a message about their community, or listen to messages left by others (Gram Vaani, 2013). A trained editor validates the recorded messages before they are shared with other callers. Such interactive systems are often operated in conjunction with community radio stations, which may then more widely broadcast the issues raised by the community in the station's local programming.

Discussion

Insights on Roles of NGOs in ICTs for Empowering Women

For NGOs to help harness the power of new ICT for rural women, they must educate their own staff about the opportunities available, the analysis for selecting/ customizing new ICT, facilitation of implementation of new ICTs, and monitoring and evaluation of the ICT implementation. This is no easy task, as there are many technologies being piloted in the development sector and many new ICT developers promoting their own products. There is significant expense involved; coupled with uncertainty, that is a deterrent to trying new ICTs. Most of the literature reports on the array of particular pilot projects, rather than assisting development practitioners in identifying strategies for selecting technology appropriate for their activities and for evaluating their impact.

Practical factors affect NGO decisions about piloting of ICT for development in rural areas. Inadequate infrastructure, such as reliable power sources and Internet and mobile phone connectivity, poses challenges to successful implementation of a new ICT (Ramachander, 2009). NGO employees are often focused on high priorities in their day-to-day development projects, rather than on exploring ICT possibilities and providing necessary support to customize and implement them with communities. They may be reluctant to pilot ICTs with great promise that are apt to be unsustainable. It is important nonetheless for NGOs to allocate some time, energy and resources to learn about and discuss new ICTs with communities, to elicit community ideas about using them and to search out solutions to challenges.

Figure 9.2: Young Women from the Community Record their Opinions for Broadcast on Alfaz-e-Mewat.

To summarize, key NGO roles can be to:

- ☆ Learn about ICT opportunities of special use and interest to women
- ☆ Share an inspiring vision of women's empowerment to staff and the community
- ☆ Network with organizations specializing in ICT use in the development sector

- ☆ Engage women in rural communities in learning about ICT options
- ☆ Identify content relevant to rural women in the area through participatory methods
- ☆ Find and secure financial and expert resources
- ☆ Help replicate suitable ICT projects for and with communities often in partnership with ICT experts
- ☆ Innovate and pilot new ICT use adaptations in consultation with communities
- ☆ Monitor and evaluate the effectiveness of specific ICTs to achieve their purposes in specific locations
- ☆ Assist with securing resources and government approvals
- ☆ Build capacity of communities to manage ICT projects locally and
- ☆ Provide platforms for discussion of ICTs and selection strategies to meet rural community needs.

Challenges

What are the challenges that NGOs and communities face in making greater use of available ICTs?

Organizational Support for Change

Stakeholders at all levels may lack the gender sensitization and ICT knowledge to inspire a desire to adopt change. It may seem hard to imagine how the ICT will really work and how to adapt it to empower women in particular (Sulaiman *et al.*, 2011). Staff and community members' interest in making communication technology changes is essential. NGOs should provide training on ICTs and network with others in the development sector working with ICTs to empower women.

Evaluating many Options

If open to considering new ICTs, NGOs have many options. It seems complicated to pick wisely from among them. Since there are many pilots of digital ICTs, and many of them quite recent, there have been few external impact assessments. How should an NGO choose ICTs to present to communities for discussion?

Quality of the Information and its Expression

Much of the discussion in this article is about the communications technologies and the suitability of the design of the ICT project for rural women. The quality, utility, timeliness and expression (local language and appropriate communication level) of the information provided for specific communities are key to success of an ICT project. To remain relevant and useful over time, the ICT must allow a dynamic process rather than static information broadcast.

Capability and Views of the Rural Women Engaged

NGOs need to interact with the women they seek to engage to determine their needs, abilities and views, and the degree of support ("human intermediation") that will be required to use the technology effectively (Sulaiman *et al.*, 2011). What NGO staff find interesting and valuable may not appeal to particular rural women.

Infrastructure Gaps

Some areas do not have reliable mobile service and/or reliable electricity service for recharging phone batteries or operating computers and the like. Many rural areas do not have Internet access.

Financial Accessibility

If women see the benefit of the new ICT, they may lack access unless it is a totally subsidized programme. How will rural women afford phones, mobile service, other equipment and service connections?

Reluctance to Partner

NGOs may be reluctant to partner with others on ICT if NGO funds are required and they are unsure of a successful outcome.

Government and other Formal Processes

An ICT such as community radio requires government approvals. Meeting all the technical requirements for approval takes time and dedication to the task. Likewise, donor-funded projects often require sophisticated evaluation and reporting that may be difficult for smaller NGOs.

Sustainability

Community radio stations face the challenge of financial viability to retain good staff and maintain and update equipment. Similarly, other ICT systems require human resources and equipment. Sustaining an ICT system requires commitment, resources and know-how.

Conclusions

To accelerate the pace of access to ICT for rural women in India, NGOs should keep these points in mind:

Knowledge Sharing

NGOs need to learn about digital ICTs, share their learning with women in rural communities, and discuss with women their priorities and their access to ICTs. NGOs can connect communities with experts and trainers to involve grassroots women in designing and implementing a project suited to women's priorities. Broadcasting generic content has some value, but it is more likely that interaction with rural women will lead to answering their specific questions and addressing their needs. Through NGO networks specializing in ICT, NGOs can identify providers, technical and content support, potential funders and successful funding strategies. There is need for study of the impact of the pilot projects to refine them and identify best practices.

Digital Empowerment Foundation catalogues and highlights the features of ICT projects in rural India, with special attention to those designed to empower women. It also shares creative solutions that are emerging to various infrastructure inadequacies.

Such platforms are essential for helping NGOs to analyze the vast array of proliferating pilot projects,

Design of ICTs for Empowering Women

Relevance of particular ICT uses to "rural women" depends on the education level and circumstances of the specific women in specific localities. In view of widespread female illiteracy, voice-based and symbol-based systems for mobile phone technology will be important for effective communication with many. In addition, NGOs should support female literacy programmes. But there are also rural women with sufficient education to engage in livelihoods related to ICTs. Rural women's involvement in the design and selection of ICT uses will bring these issues to the forefront.

The Centre for Research on Innovation and Science Policy in Hyderabad, India, in collaboration with partners in Bangalore and Scotland, has developed a three-stage consultative approach for designing programmes for women (Centre for Research on Innovation and Science Policy, 2013). This approach includes needs assessment, women's aspirations, available resources, pros and cons of intervention options, and opportunities for convergence.

Community Networks

Women's empowerment through ICTs builds on the existence of networks with which women can engage and gain confidence and skills (Sulaiman *et al.*, 2011). Self–Employed Women's Association (SEWA) in India and other self help groups provide a sound support network for women, who grow in their ability to make decisions, solve problems, and plan for their futures. The technology itself does not empower an individual woman without more; it is the participation in a learning and problem-solving community that supports her confidence and personal agency. ICT broadcasts can spark discussion among local radio listener and television viewer groups (Sulaiman, 2011). NGO support may be required to facilitate local level communications in initial stages.

Affordability

To enable access, women below poverty level will need government subsidies or private sponsorships through foundations or other donors for mobile phones/service. Alternatively, ICT uses that are tied to livelihoods may become a cost of doing business for small farmers and entrepreneurs. Community kiosks located in spaces accessible to women are another way to make new ICT available. Government, community contribution, and/or the private sector should fund them.

Emergent Nature of ICT Technology

There are many pilot ICT programmes. NGOs should monitor the experience of pilot programmes, address problems and refine the models. NGOs should study the impact of ICTs for rural women on development outcomes and share these results within the development sector and with policymakers. The underlying technologies are rapidly changing, and we can expect a high learning curve, ongoing need for evaluation and constant refinement. We can expect that pilots of various kinds will continue for years and that a range of good models will be developed and be customized for local needs.

NGOs should be vital agents for bringing ICT into development sector work to empower rural women. Rural women lag far behind in the education and information access necessary to make good decisions for their own well-being, that of their families and that of their communities. While there are challenges and complex issues to address to tap the potential of new ICTs for women's empowerment, it is clear that NGOs are essential partners for communities to successfully seize the opportunities that ICTs can offer.

Acknowledgements

I am grateful to Jaypee University of Information Technology, Solan, H.P., the NAM Science and Technology Center, Delhi, and Read India for the invitation to participate in the conference that has led to production of this book. I thank the Sehgal Foundation Communications team for sharing its ICT experience and photographs, the many community media contacts who have educated me during Sehgal Foundation and UNESCO's annual Rural Voices conferences, and Padmavathi S for assistance with the formatting of this paper.

References

1. Barefoot College, 2013. Disrupting poverty: How Barefoot College is empowering women through peer-to-peer learning and technology. Report. http://www.barefootcollege.org/disrupting-poverty-how-barefoot-college-is-empowering-women/ (accessed 28 May 2013).
2. Beyond Access, 2012. Empowering Women and Girls at Libraries. Issue Brief, Beyond Access Libraries Powering Development. http://beyondaccess.net/ (accessed 19 Jul. 2013).
3. Cecchini, S. and Scott, C., 2003. Can Information and Communications Technology Applications Contribute to Poverty Reduction? Lessons from India. *In:* Information Technology for Development 10, pp. 73-84.Centre for Research on Innovation and Science Policy, 2013. How to Develop a Project/Programme for Rural Women. http://www.reachingruralwomen.org/stages.htm (accessed 19 Jul. 2013).
4. CGNet Swara, 2013. Webpages "About" and "Recent Reports". http://www.cgnetswara.org (accessed 28 May 2013).
5. Department of School Education, Haryana, 2013. Announcement at Sehgal Foundation by Smt. Surina Rajan, Principal Secretary, Haryana Department of School Education on 3 Apr. 2013 of toll-free Haryana Right to Education helpline number, 180030100110.
6. Development Alternatives Group, 2013. Webpages on TARA Akshar Past Centres and Current Centres, http://taraakshar.com/TaPastCentres.html and http://taraakshar.com/TaCurrentCentres.html (accessed 12 Jul. 2013).

7. Digital Empowerment Foundation, 2013a. Chanderi Weavers ICT Resource Centre. http://defindia.net/cwirc/ (accessed 11 Jul. 2013).

8. Digital Empowerment Foundation, 2013b. Report on "Smart card project helps the unbanked poor, reduces fraud". http://defindia.net/2012/06/06/smart-card-project-helps-the-unbanked-poor-reduces-fraud/(accessed 11 Jul. 2013).

9. Digital Empowerment Foundation and UNICEF, 2013. Use of Mobile Phones for Social and Behaviour Change. Consultation paper (draft May 8, 2013), India, p. 10-23. http://defindia.net/2013/05/14/use-of-mobile-phones-for-social-behaviour-change/ (viewed 18 May 2013).

10. Government of India, 2008. E-Governance and Best Practices. Report of the Sixth Central Pay Commission, Annex 6.3.1, pp. 256-269. http://digitalknowledgecentre.in/files/2012/02/e-Governance-and-best-practices.pdf (accessed 14 May 2013).

11. Gram Vaani, 2013. Webpages on Gram Vaani and on Jharkhand Mobile Vaani, India. http://www.gramvaani.org/?page_id=343 and http://www.gramvaani.org/?p=495 (accessed 19 Jul. 2013).

12. Gulati, A., 2010. Using ICTs to M-power Women's Self Help Groups in Rural India, pp. 2-4, 8-9. http://papers.ssrn.com/sol3/papers.cfm?abstract_id=1715237 (accessed 19 May 2013).

13. Gyandoot, 2013. Webpages on Gyandoot intranet, Dhar District, Madhya Pradesh, India. http://www.gyandoot.nic.in/ and http://dhar.nic.in/GYANDOOT.htm (accessed 14 May 2013).

14. Indian Institute of Management, Ahmedabad, 2002. Gyandoot – Rural Cybercafes on Intranet A Cost-Benefit Analysis. Report Centre for Electronic Governance. http://www.iimahd.ernet.in/egov/documents/gyandoot-evaluation.pdf (accessed 19 Jul. 2013).

15. Jorge, S., 2010. Seven policy tips to ensure rural women equal access to ICTs. Article *In:* GenARDIS 2002-2010: Small Grants that made Big Changes for Women In Agriculture. GenARDIS is Gender, Agriculture and Rural Development in the Information Society, Canada, pp. 37-38. http://genardis.apcwomen.org/en/node/151 (accessed 5 Jun 2013).

16. Khalafzai, A. K., Nirupama, N., 2011. Building Resilient Communities through Empowering Women with Information and Communication Technologies: A Pakistan Case Study. Sustainability, 3 (1), pp. 82-96. http://www.mdpi.com/2071-1050/3/1/82/pdfý (accessed 11 Jul. 2013).

17. Kumar, T., 2009. E-Sanchar (e-Speech Application through Network for Communication, Help and Response), 13th National Conference on e-Governance, India, p. 2. http://indiagovernance.gov.in/files/E-sanchar.pdf) http://indiagovernance.gov.in/files/E-sanchar.pdf (accessed 11 Jul. 2013).

18. Mehta, P. K., 2010. Role of Information and Communication Technologies on Improving Livelihoods: A Case of Mewat. Working Paper No. 1, Institute of Rural Research and Development (An Initiative of S.M. Sehgal Foundation) India, pp.

1-12. http://www.smsfoundation.org/pdf/Role per cent 20of per cent 20ICT per cent 20on per cent 20Livelihoods- per cent 20A per cent 20Case per cent 20of per cent 20Mewat.pdf (accessed 19 Jul. 2013).

19. *Mera Swasthya – Meri Awaaz,* 2013. Website *Mera Swasthya- Meri Aawaz* (My Health – My Voice), Uttar Pradesh, India. http://meraswasthyameriaawaz.org/main (accessed 19 Jul. 2013).

20. Ministry of Information and Broadcasting, 2013. Community Radio Stations at a Glance. Government of India. Report on website. http://www.mib.nic.in/linksthird.aspx (accessed 19 Jul. 2013).

21. Nath, V., 2001. Empowerment and Governance through Information and Communication Technologies: Women's Perspective. International Information and Library Review, 33, pp. 317-339.

22. National Portal of India, 2013. Website of the Government of India. http://www.india.gov.in (accessed 19 Jul. 2013).

23. NextDrop, 2013. Website page "About Us", India. http://www.nextdrop.org. (accessed 19 Jul. 2013).

24. Ramachander, S., 2009. Implementation and Outcomes: Evidence from Information Kiosks in Rural India. Fourth Conference on Communication Policy Reseach, South, Negumbo, Sri Lanka, pp. 12-13. http://www.cprsouth.org/ (accessed 19 Jul. 2013).

25. Rizvi, S.M.H., 2011. Lifelines: Livelihood Solutions through Mobile Technology in India. *In:* Grisham, D. J., Shalini, K., Strengthening Rural Livelihoods: The impact of Information and Communication Technologies in Asia. Practical Action Publication Limited, UK and The International Development Research Centre, pp. 53-70.

26. S.M. Sehgal Foundation. 2007. Press release, "Fulfilling the Dream of Literate Mewat". http://www.smsfoundation.org/Fulfilling per cent 20The per cent 20Dream.html (accessed 12 Jul. 2013).

27. Spaven, Patrick, 2011, Evaluation of the Commonwealth of Learning, Brighton, England, p 46. http://dspace.col.org/bitstream/123456789/444/1/ExternalEvaluation_2009-2012.pdf (accessed 12 Jul. 2013).

28. Sulaiman V, R., Kalaivani, N.J., Mittal, N., 2011. ICTs and Empowerment of Indian Rural Women. Working Paper 2011-001, Centre for Research on Innovation and Science Policy (CRISP). http://www.crispindia.org/docs/4 per cent 20CRISP per cent 20Working per cent 20Paper-ICTs per cent 20and per cent 20Empowerment per cent 20of per cent 20Women.pdf, (accessed 8 Jul. 2013).

29. United Nations, 2005, Gender equality and empowerment of women through ICT, Women 2000 and Beyond. Report, Division for the Advancement of Women, Department of Economic and Social Affairs, United Nations, New York, NY, USA, p. 8. http://www.un.org/womenwatch/daw/public/w2000-09.05-ict-e.pdf (accessed 14 Jul. 2013).

30. Vidiyal, 2013. Website of Vidiyal, in conjunction with the Commonwealth of Learning, Canada. http://vidiyalngdo.in/lifelong.htm (accessed 28 May 2013).

31. Vital Waves Consulting, 2011. Women and Mobile: A Global Opportunity. Report, U.K., p 8. http://www.vitalwaveconsulting.com/pdf/2011/Women-Mobile.pdf (accessed 28 May 2013).

32. Vodafone India Foundation and Digital Empowerment Foundation, 2011. Mahila Shakti *In:* emPowering WOMEN Through MOBILE, India, p. 13. http://mobilesforgoodwinaward.manthanaward.org/files/2011/10/emPowering per cent 20women per cent 20through per cent 20mobile.pdf (accessed 11 Jul. 2013).

33. Voluntary Association for People Service, 2011. Enhancing Women Empowerment through Information and Communication Technology, pp. 62-63. Report to Department of Women and Child Development, Ministry of Human Resource Development, Government of India by Voluntary Association for People Service, Madurai, Tamil Nadu, India. http://wcd.nic.in/research/ict-reporttn.pdf (accessed 12 Jul. 2013).

34. Water and Sanitation Programme, 2012. You Manage What You Measure: Using Mobile Phones to Strengthen Outcome Monitoring in Rural Sanitation. http://www.wsp.org/global-initiatives/global-scaling-sanitation-project (accessed 12 Jul. 2013).

Chapter 10

Women's Empowerment Gap and Sustainable Development in Sudan

Hanan Mohamed A. Karim Abbas

CSR and Sustainable Development Advisor,
Global Compact Association,
Businessmen and Employees Federation, Sudan
E-mail: hanan_abbas@hotmail.com

ABSTRACT

Sudan with its large area of almost 2,000,000 square kilometres and multicultural diversity is least worried about women inequality and the inability of uneducated women to send their children to schools. Women are strongly positioned in all sectors in Sudan, from presidential advisors, judge, to the lower levels of labourers. Women constitute more than 57 per cent presence at all work force levels.

The Sudan women desperately need empowerment in other major issues. A huge gap exist in reproductive health and Female genital mutilation (FGM), environmental and agricultural involvement of women, family income generation and financing projects, and enabling cultural change.

As a matter of fact, the three pillars of society Government, Private Sector and Society, have failed to coordinate works in order to achieve any significant results in women empowerment for a sustainable economy.

Private Sectors, in particular, corporate social responsibility (CSR) policies are still dominated by charity and Philanthropy. Telecommunication companies, in particular, give out millions of dollars annually to the Philanthropy project, which then can be directed to great and creative CSR policies to benefit families in the vast Sudan. Other countries report many success stories by using the power of women in using mobiles and networking.

Keywords: *Enrich women's lives for a sustainable development by ICT.*

Introduction

The reflection of Sudan until 2011 was the largest country in Africa, and was known as the million square mile country. In 2011 South of Sudan was separated and at present Sudan stands as the third largest country in Africa, with around 6,000 ethnic groups with different local languages and dialects. Islam and Christianity are the dominating religions in the country in addition to other African traditions. Sudan is surrounded by seven neighbours, each one with its own political and economical problems that affects Sudan both, directly and indirectly.

Civil war emerged as a north-south conflict, but recently has spread to the other zones of Darfur and the Nuba mountains in the West and to Eastern Sudan. This war has resulted in death, and the displacement of 4-5 million people. Many refugees now live in the neighbouring countries. In addition, two million internally displaced persons live around the capital,the greater Khartoum.

Most of the Sudanese are living below the poverty line. There are approximately 720 active NGOs in Sudan including and not limited to national NGOs, INGOs, UN agencies, and CBOs. This large population, work mostly in the field of humanitarian assistance while very few are involved in any developmental projects in Sudan.

The presence of significant natural resources such as reserves in oil, gold and other precious minerals and metals, makes the country attractive to many investors.

Basic Facts on Sudan

- ☆ Population: 34,847,910
- ☆ The country is divided into 17 states
- ☆ The Legal System comprises of: mixed of Islamic law and English common law
- ☆ Religions: Sunni Muslim, small Christian minority mainly Protestants
- ☆ Sudan is rich in natural resources namely: petroleum, gold, hydropower, and small reserves of iron ore, copper, chromium ore, zinc, tungsten, mica, silver
- ☆ Land use: arable land 6.76 per cent, permanent crops 0.07 per cent, others 93.17 per cent (2011)
- ☆ Irrigated land: 18,900 sq km (2010)

Materials and Methods

This paper is part of an attempt for a comprehensive look at a different approach towards poverty reduction for sustainable development.

The first hand information is collected by field research in the troubled areas of Sudan. This involved extensive travel all across Sudan, living with the local people and participating in their lives. The largest sample of women taken was 25 and our total population was approximately 500. Group discussions as well as brain storming with these women and their community leaders are the main source of information.

A different group discussion with activists and concerned persons in the field was conducted regularly. Formal and informal discussions were held especially with displaced women and nomads. The activists were diversified to include males and females in different sectors and professions *i.e.* academia, health and nutrition, law, clergies, civil society (local and international), UN Agencies, administration, anthropologists, historians, government authorities and the business sectors.

Secondary information is also derived from researchers that have already published their studies by using the United Nation sources, and also from undergraduate researches which are mainly from the Ahfad University for Women.

Results and Discussion

Women of Sudan with Change in Gender Norm: Injustice, with a Semi State of Gender Equality

Evidence had shown that the influence of gender norms in the Sudanese lives are much relaxed as allowing more inclusive development sharing and sharing new responsibility. Women in Sudan with their different education levels, work to support families without comprising their family obligations. The wars, conflicts, migration and immigration have resulted in 59 per cent of women-headed households of a family made up between 7 to 15 persons.

The inequality of women is actually less worrying. In spite of the fact that minor feminist might see it different but as a society it has been proven that the Sudanese culture has the maturity in appreciating the role of women in society and in work place. There are few odds, however, which can be dealt in a systematic manner.

Many of highly educated females have no gender issues when it comes to life's activities. They have great sense of satisfaction to leave the father or the husband takes care of the major family decisions, they are happy to make family meals without any sense of inequality or injustice.

The following facts and figures are highlighted based on statistics collected over a period of seven months of November 2012 – May 2013:

- ✰ Women population in Sudan is approximately 50 per cent of the population
- ✰ Women hold 26 per cent seats in the National Parliament
- ✰ Women were elected to 26 per cent of positions in the lower house and 11 per cent of positions in the upper house
- ✰ Sudan has the first female High Court Judge in African and the Mena Region
- ✰ Sudan has women markets
- ✰ Women and men have the same right to vote and to stand for election
- ✰ Unemployment among new female graduates is estimated at only 35 per cent
- ✰ The presence of women as a percentage of work force per category:
    - 100 per cent midwifery
    - 97 per cent of tea sellers

- 96 per cent of working from home
- 70 per cent in nursing
- 67 per cent in academia and research, however very few women headed an academic or research unit
- 60 per cent in community, social, and personal services
- 58 per cent in medical field
- 57 per cent in middle management
- 57 per cent in journalism
- 56 per cent in retail trade in fashion and related industry, while 2 per cent in market bases for retailing
- 56 per cent in government offices
- 55 per cent in banks, financing and insurance
- 45 per cent in telecommunications, while 3 per cent in related job fields *i.e.* construction of satellites
- 45 per cent in supervisory positions
- 38 per cent in industry
- 35 per cent in restaurants and hotels
- 33 per cent in architecture
- 30 per cent in customs
- 17 per cent in private businesses (entrepreneurs)
- 15 per cent in wholesale trading
- 15 per cent in real estate, and business services
- 15 per cent in donkey ridden transport in rural areas
- 10 per cent in top management
- 7 per cent in water services
- 6 per cent in construction (not manual and labour workers)
- 5 per cent in electricity production, but 0 per cent in hydro electricity production (field sites)
- 4 per cent in oil field production, while 40 per cent in administration
- 3 per cent in field work of mining, while 35 per cent in mining administration
- 2 per cent in transport and storage
- 2 per cent in traffic police force
- 1 per cent in police force, 30 per cent in administrative posts
- 0.3 per cent in the army, 20 per cent in administrative posts
- 0 per cent in public transport vehicle drivers
- 0 per cent in all in physical labour or "strength" jobs *e.g.* army combat, fire fighting, etc.

Positive Discrimination in the Work Place in Favour of Women

- ☆ Women are given priority in microfinance programmes. The concern is that there are limited women who apply for microfinance because they are not risk takers.
- ☆ Women are not allowed to work between 10pm and 6am with the exception of women in administrative, professional, technical work or health services.
- ☆ The divorce dowry is the property of the wife's family.
- ☆ It is observed and noted that at the work place women and men are treated equally. In many instances men complain that management favours women.
- ☆ Women have a legal right to a paid maternity leave of 8 weeks and afterwards, women are aloud of an hour release from work per day for breastfeeding. Women also have the privilege of three months extension of maternity leave without being paid.
- ☆ Women are given a day off from the work every month in recognition of the inconvenience associated with the first day of the menstrual cycle. This privilege continues even for women on menopause.

Where the Injustice to Women Comes in Sudan?

- ☆ Women are left to take care of the family while the husbands are either in war, migrated, immigrated or mining for gold
- ☆ Women in Sudan with their different backgrounds of work have to support for their families without compromising their family obligations and with little facilities in hand
- ☆ 59 per cent of women are headed households
- ☆ Women have restricted access to land ownership
- ☆ Women have limited access to bank loans and other forms of credit and finance, except in the case of micro-finance
- ☆ Women cannot travel abroad without the permission of their husbands, male guardians. However, this prohibition is not strictly enforced.
- ☆ Women can only use contraceptives after the permission of the husband.
- ☆ Both husband and wife could ask for divorce, but obtaining it is much difficult for women than for men.
- ☆ A husband has the right to divorce his wife unilaterally, without turning to the court, by merely saying "I divorce you"
- ☆ Women in conflict areas like all women in such difficult situations are vulnerable and suffer most; facing rape, coercion, sexual abuse, psychological harm, arbitrary deprivation of liberty, forced for prostitution, etc. In many conflict areas in Sudan, men and young boys, to a limited extend, suffers the same problems.
- ☆ 60 per cent of women in war zones are displaced
- ☆ Limited law and regulation are in place for supporting women
- ☆ There is no national policy for women empowerment

In addition to the above problems, there are crimes imposed by women on their sisters and daughters. These include and are not limited to the prevalence of:

- Female genital mutilation (FGM) or female genital cutting (FGC)
- Forced marriage at an early age
- Polygamous unions
- Deliveries unattended by trained personnel
- Resistance for the use of contraceptives or any form of protection

Statistics also showed that:

- 82 per cent of females and 81 per cent of males agreed that 'a woman should tolerate violence in order to keep her family together'
- 68 per cent of females and 63 per cent of males agreed that 'there are times when a woman deserves to be beaten'
- 47 per cent of women and more than 37 per cent of males agreed that 'it is okay for a man to hit his wife if she will not agree to have sex with him.'

The Taboo of Reproductive Health and FGC/FGM

"When conflict is extended, public health activities, including immunization and surveillance systems, can be substantially reduced, dismantled and destroyed, as happened during recent conflicts in Rwanda, Sudan, Liberia, Chechnya, and Iraq." A quote by Hynes, H. Patricia. On the Battlefield of Women's Bodies: An Overview of the Harm of War to Women." *Women's Studies International Forum* 27 (2004): 431-45.

"No matter how good their intentions may be, the reality is that an early marriage generally offers no protection at all – in fact, the opposite is generally true – and it strips many young girls of their childhood, their dreams, their basic human rights and their health". Early Marriage. *International Women's Health Programm*e, The Society of Obstetricians and Gynaecologists of Canada.

Basic Data for Sudan

- The estimated prevalence of FGM in girls and women within the age of 15-49 years is 90 per cent (under gone some form of FGC)
- The prevalence of FGM in girls and women within the age of 15-19 is estimated at 85.5 per cent
- 36 per cent of women have married before the age of 18
- 12 per cent have married before the age of 15
- 24.7 per cent of girls at the age of 15-19 have married or are in union
- 27.5 per cent of women are in polygamous unions
- Maternal mortality ratio is 750 per 100,000 live births, it is the largest reason for death after malaria
- Teenagers mortality ratio is 800 per 100,000 live births
- 1 of 20 women died or suffer for long lasting injuries or disabilities, *i.e.* obstetric fistula, uterine prolapsed infertility and depression.

- Infant mortality rate is 70 per 100,000 live births
- Neonatal Mortality rate is 31 per 1000 live births
- Percentage of deliveries attended by trained personnel is 57 per cent
- Accessibility to emergency obstetric care is low as indicated by caesarean section rate of 3 per cent
- Use of contraceptive prevalence rate is 7 per cent
- Total fertility rate (births per women) is 4.05 per cent
- HIV/AIDS – adult prevalence rate is 1.1 per cent (2009 estimates)
- HIV AIDS – people living with HIV/AIDS is 260,000 (2009 estimates)
- HIV/AIDS – deaths is 12,000 per annum (2009 estimates)

Based on the above data, there is a great urgency to accelerate progress towards a national programme with the following priorities:

- Maternal and neonatal health
- Family planning
- HIV/AIDS prevention
- Harmful traditional practices
- Adolescents and youth reproductive health
- Infertility
- Screening for breast cancer and cervical cancer
- Management of menopause

Using the Experience of Others, How Can We Achieve Priorities?

The priorities can be implemented as a comprehensive strategy that would involve the whole community and not limited to clergies (Masheikh/imams) and community leaders. The activities should include:

- Sending girls to schools and colleagues
- Teach the girls a useful handcraft or skill
- Avoid early marriage
- Prevent forced marriages
- Stop FGC/FGM
- Stop women delivering babies at home with assistance by a non-qualified midwife
- Break the taboo for women and men to speak to the specialist about sexual and reproductive health issue
- Give the parents the right for family planning using contraceptives or other means
- Encourage periodical preventive check-up for breast and cervical cancer
- Encourage parents to provide their children with basic reproductive education

Women Empowerment for Income Generation

In spite of Sudan being rich in oil and gold, farming is still the main source of income to the majority of people. Farmers in general have difficulties but women farmers specifically have overlapping difficulties, such as: poverty, discrimination policies and environmental degradation of land. Women can hardly cultivate enough to feed their families. Climate changes make the situation worse, as it brings in droughts and floods to destroy crops that have been financed with great difficulties. Global climate change ruins the crop and has made farmers with traditional knowledge obsolete, resulting in many abandoning their farming way of life.

The farming inputs provided by the local authorities many times exclude women and deny them any seeds or fertilizers. They tend to forget that 59 per cent of families are women headed and women have to secure their own credit and other farming inputs.

Women farmers who are the majority would need tools, resources and technical knowhow. The solution is to create a small number of female entrepreneurs that have minimum level of education, and provide them with basic support in order to establish a small scale of ICT business. Besides the selling of ICT enabled online services, the business would then provide other women with seeds and other farming supplies. Through the ICT centre messages can then be sent periodically to remind the farmers of important farming dates, *e.g.* period of using fertilizers and spraying the crop, etc. This approach has proved to be effective as it provides the local farmers with sustainable food supplies and generate a stable income for the families.

The ICT entrepreneurs could run these small businesses as a community centre and also to provide other services:

- ☆ Education programmes that promote literacy and computing skills, along with access to the facilities of a small ICT centre.
- ☆ Money transfer using mobile telephones *i.e.* payment to suppliers for seeds, fertilizers, etc.
- ☆ Trainings on reproductive health, HIV/AIDS prevention, peace building and political participation.
- ☆ Trainings to help women farmers adapt to climate change by teaching them about changeable weather patterns, providing instruction for adjusting soil preparation, planting and harvesting accordingly, and how to harvest rain water and dig shallow wells in villages.
- ☆ The centre would also be a data base for the various activities:
  - Informal job providers for women where women would be accessible by a call. This would enlarge the sector of "work for cash".
  - Income generating projects that have succeeded in other areas of Sudan:
    1. Growing crops that have high demand with minimum risks *e.g.* the Moringa Oleifera trees.
    2. The production of "Home Made Jams" from the locally grown fruits *e.g.* oranges, grapefruits, mango, guava fruit,etc.

3. Biogas and efficient use of agricultural residues/waste.
4. Improving the production of Sudan local cheeses (jebnabaydah and jebnamadafarrah).
5. Improved methods for drying food like fish (locally known as fasekh), herbs and vegetables such as onions.
6. Introducing Small Poultry Production Unit for small farmers.
7. Introduction of new simple methods for collecting and grinding herbs.

This would mean that Women

- ✰ Gain the resources they need to grow and produce food, alleviate hunger, improve health and nutrition, and fuel growth of local economies.
- ✰ Work together to grow crops, build a network of women farmers who could share resources and boost their economic status, use elderly women to transmit skills and lessons to younger women.
- ✰ Improve their knowledge and know how on the environment that would increase their agricultural yields, enable them to boost food security and generate income from surplus crops.
- ✰ Share to expand their centre to include literacy classes and to provide other needed services to the community, *i.e.* electricity and diging of wells.
- ✰ Extend knowledge to young female farmers and encourage them to enrol in agriculture knowledge and education programmes.
- ✰ Use part of their income to pay for the education of their daughters.

Introducing ICT

Worldwide people recognise the importance of Information and Communication Technology (ICT) in sustainable development. It is an effective tool that makes access to all location without boundaries. Telecommunication, radio and television bring about unprecedented changes in the way people communicate, conduct business, pleasure, make social interaction, and carry out massive humanitarian and developmental work. They are an effective media given the vastness of Sudan with the problems of poor means of transportation, lack of paved roads, and expensive travelling.

ICT have made peoples' lives better especially in remote areas. People are simply reachable and any type of information is accessible at a finger tip:

1. Medical consultation
2. Connecting buyers to sellers
3. Connecting sellers to markets
4. Match making between sellers and buyers
5. Transferring information and increasing awareness from how to farm and cultivate crops to simple meal recipes using local ingredients
6. Money transfers

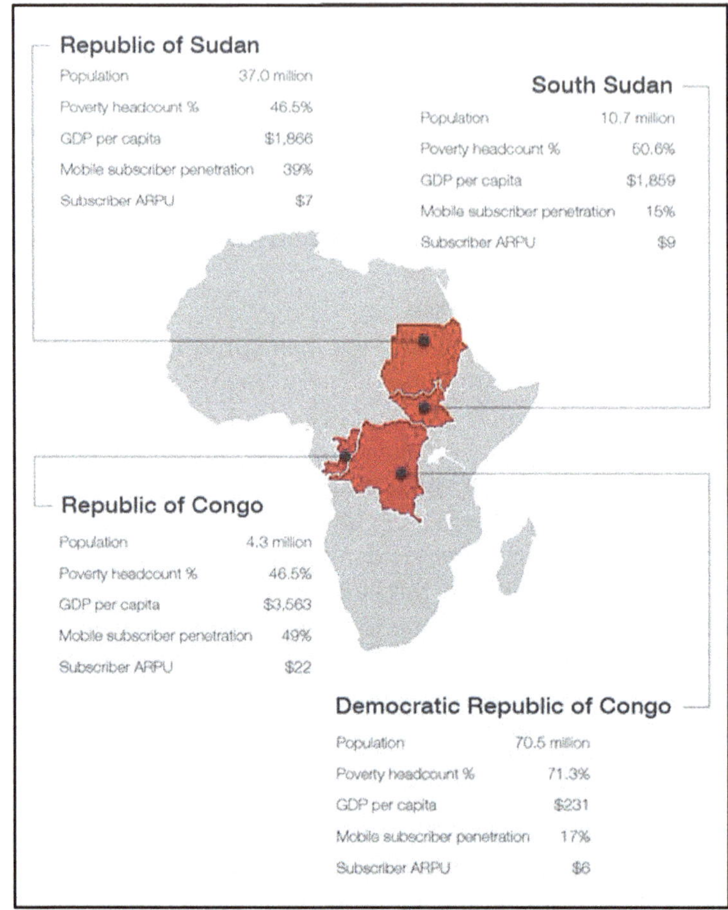

Figure 10.1: Comparison of the Sudan Mobiles Subscribers in Comparison to Neighbouring Countries.

***Source*: Wireless Intelligence (ARPU, penetration, Q4 2012), UN (Population, 2012), World Bank (GDP, poverty headcount, latest available data).**

Less than a quarter of the population had subscribed to a mobile service by the end of 2012. There is no firm statistics indicating that fewer women than men owned a mobile phone in low- to middle-income families. There is however, a statistic that shows that women talk time is 30 per cent more than men. However, based on the survey made, women are the poorest communities that prefer to use money for the family rather than invest in mobile phones. On the contrary, men consider that buying mobile phones is a valuable investment.

This clearly indicates the window of opportunity open to approximately 7 million Sudanese women without access to the potentially life-enhancing tools. By overlooking the women's market, the mobile phone industry could have estimated revenue opportunity of over $1.5 million.

This gap also suggests the lost opportunity for women, particularly the resource-poor women, who would gain most from the mobile phones' potential benefits. A more productive and sustainable use of ICT is an identified need in low income women in remote areas.

ICT Women Programme developed/promoted aims to increase the access of women to and use of mobile phones and life-enhancing mobile services. The ICT access is made available to poor women in remote areas without confining its use to the high class strata in society.

ICT in Sudan had already achieved significant success with the needy communities without affecting the service providers' bottom line *i.e.* still making profits. In spite of the exiting challenges, ICT succeeded in:

1. Cost saving and reducing road hazard.
2. Mobile phones offer women security and peace of mind, such as the ability to call a doctor during a time of need when hospitals are not nearby.
3. Mobile access led 93 per cent of women to feel safer and 85 per cent to feel more independent.
4. Both skilled and unskilled persons are reachable. The use of mobile phones increased the "work for cash" populations of causal and informal labors.

Increasing Demand on ICT for Sustainability

ICT is needed for greater cause, such as:

1. Accessing and reaching nomads that are isolated and scattered, in order to:
   a. Put them on the map
   b. Cluster them in order to provide services such as: schools, clinics, training and different awareness programmes, etc.
   c. Protect them
2. Empowering women in order to assist them in:
   a. Reproductive health and FGM/FGC
   b. Environmental issues
   c. Agricultural involvement and potential
   d. Family income generation and financing projects
   e. Culture change

In the above two points are referred to women who are heading and supporting between 7-15 family members without a husband. The husband might go for gold mining, migrated, immigrated or fighting in one of many conflict areas of Sudan.

The telecommunication sector is the fastest moving industry in Sudan with significant infrastructure and reliable networks. Considering this advantage against a vast country that has few of roads and no railways, it is found that mobile phones and communication technology are the easy and fast way to reach people in remote nomadic areas. These people livein condition of low income, low level empowerment and are isolated socially, with minimum livelihood facilities of water, schools and health.

Sudan Telecommunication Providers

The Sudan Telecommunication Providers can be seen in Table 10.1.

Table 10.1: Existing Operators in Sudan. Data up to Mach 2013.

Operator/Service Provider of Mobiles and Landlines	*Date of Licence TECHNOLOGY*	*Country Wide Coverage*	*Subscribers*
Zain Sudan	14-Aug-96	120 locations	14,000,000
MTN Sudan	25-Oct-03	35 locations	2,700,000
Sudani One	02-Feb-06	145 locations	4,000,000
Canar Telecommunications/ Etisalat	Apr-05	Unknown	1,000,000

The above telecommunication companies provide simple operation of ICT: voice, data and video communication. The services can provide the end-users in remote areas continuously, which are for life saving and others to contribute significantly to sustainable development. The services cover:

- ✰ Provide information that would link civil society sustainability projects with private sectors (generally as donors).
- ✰ Link individual donors with NGOs/CBOs working at the community levels.
- ✰ Awareness messages to the society needed in different fields.
- ✰ Transmitting information on climate changes and their effect of crops and farms.
- ✰ Transmitting health and safety programmes.
- ✰ Spreading knowledge on income generation projects.
- ✰ Agricultural advice and information.
- ✰ Culture change initiatives.
- ✰ Environmental protection programmes.
- ✰ Link small producers with market.
- ✰ Accessibility to market information.
- ✰ Providing a traceability system for small producers.
- ✰ Benefitting farmers by reducing middle men and brokers.
- ✰ Providing sellers with the update prices at each remote location.
- ✰ Providing financial support for the unbanked sectors of the society.
- ✰ Assisting in money exchange and transfer.
- ✰ Transferring knowledge to farmers as to advance their farming skills and techniques that help them to improve productivity, reduce costs, and increase income.
- ✰ Young entrepreneur creation who can run communication centres.

The above services would assist women in their struggle to overcome various obstacles related to the culture and traditions in Sudan. The idea is to empower and encourage women to use and get the maximum benefit from the continuously developing world of ICT. Yet, without having a clear vision of how to overcome the challenges hindering the Sudanese women, it will be impossible to bridge the digital gap.

For telecommunication service providers, many challenges are yet to be overcome. Zain, being among the largest providers, explained that the vast geographical landscape of Sudan makes it difficult for operator companies to roll-out mobile networks across the entire territories, and in many areas (like South Kordofan). The residents are nomad Arabs who move from one area to another with their cattle. Hence it is challenging to provide services of mobile phones to such users as well as to initiate corporate social responsibility projects for them.

For Sudanese women the challenges of ICT would be:

1. Cost of devise plus the running cost.
2. Provision of electricity. Only 40 per cent of the population have access to electricity.
3. Usage of equipment and how to maintain it.
4. Coping with illiteracy. In many areas women and girls have limited access to education and may be illiterate, so that complicated navigation of basic SMS features might be a problem.
5. Reluctant to use mobile phones. 22 per cent of resource-poor women did not want mobile phones because they do not know to use it.
6. Limited network coverage in remote areas.

Towards a National Strategy

Based on the above challenges, the recommendation would be launching an ICT National Strategy with the objective of empowering women for sustainable development. The trial would be to have a comprehensive national strategy with all major players including and not limited to the government sector. The ideal players are actually the private sector, civil society and the unions to encourage the governments to lead more initiatives at the local level. The initiatives include further allocation of human and financial resources, and paving the way for ICT service providers for further investments in the communication infrastructure. This will ultimately give mutual benefit to all sectors.

Accordingly, the effective role of women in sustainability can only be achieved through the equilibrium of the three pillars:

1. The Government
2. The Private Sector
3. The Civil Society

In any society, the three pillars function together in order to achieve the desired sustainable development for the benefit of people and its community. When this condition is observed a state of equilibrium is maintained for the benefit of all. Accordingly, the effective role of women in sustainability can only be achieved through the equilibrium of the three pillars.

The three pillars can then effectively engage through:

1. The power of Law, legislation and good governance
2. Responsible media
3. Donors

When the roles are overlapping and the duties and responsibilities are not observed by the three pillars, a state of disequilibrium occurs, which will be very challenging to correct and it affects the poor and the needy the most.

In Sudan a state of disequilibrium is brought about by:

- ✰ The government owned profit making companies
- ✰ The private sector built schools and provide health care
- ✰ The civil society has a long list of political agenda

Equilibrium had been further jeopardized by:

- ✰ Lack of coordination between the pillars
- ✰ Lack of commitment from the Government
- ✰ Civil Society is preoccupied by humanitarian activates and long political agendas
- ✰ US Sanctions prevent donors from providing developmental support
- ✰ Private Sector concentrates in philanthropy and totally ignores Sustainable development
- ✰ Government is preoccupied by conflicts
- ✰ Lack of good governance in place
- ✰ Media is dominated by irresponsible journalists
- ✰ Spread of corruption

The effective of ICT National Strategy can be formulated into two stages:

1. The Commitment of Government and civil society in action to participate, support and actively contribute to the National Strategy in its extended dimension.
   a. Create a National Body to formulate the strategies
   b. Ensure the right laws and regulations are in place
   c. Ensure that laws and regulations have the mechanism of implementation and follow-up
   d. Assist the private sector to invest further in ICT
   e. Pave the way for new ICT investors.
2. For immediate action, sufficient budget should be diverted from the CSR and sustainable development budget of private sectors, mainly from the

communication and telecommunication companies. The resources should reach out for the programmes that are already in the pipeline and the programmes developed and successful in other countries similar to Sudan.

a. Attract CSR and sustainable development fund from the region, *e.g.* GSMA mWomen Programme of Etisalat Dubai.
b. Encourage CBOs, NGOs, INGOs and UN Agencies to take up CSR projects of ICT in nature for women empowerment from service providers such as: Zain, Sudani, MTN and Canar (Etisalat).
c. ICT service providers can develop partnership with other companies, foundations and civil society organizations, in helping millions of women to have access to life skills, education, job training and opportunities critical to their success.
d. Encourage donors to divert part of their funds towards the development of ICT projects.
e. In partnership with international institutions, local authorities and operators, ICT should reduce the digital divide and promote universal access to information and communication technologies.
f. Intensive usage of radio and television to spread the awareness of the importance of ICT for the development of women especially in rural areas. On the other hand, newspapers can be used in the area of higher degree of literacy.

Conclusions

In the country that is rich in mineral, oil, land and water resources, people in Sudan should expect prosperity and a dignified standard of living. Sudan has women as 50 per cent of its total population, they are generally stranded and powerless in the system that lack the basic rights to education, health, water and decent living. The remaining 50 per cent of population is men, which are either frustrated, or immigrate, migrate, or go for gold mining, or fighting in one of conflict areas of Sudan.

The women are actually the hope of the nation to bring about a change. Empowering women is the only feasible way to take the country out to a real sustainable development. Their intelligence, charisma, power and ethics to protect their families and serve their communities should be invested in the most positive and constructive manners.

The present gap is created by the preoccupation of the Government in conflicts of forgoing the need of basic services, the restricted humanitarian aid provided by the donors due to the USA sanctions, the corruption within the system, and the limited support extended by the private sectors in the form of philanthropy.

The use of ICT to empower women in remote and scattered areas, are proved gradually as to be the best and fastest ways to achieve results. The ability to learn and adapt to what is good for them, their families and county will render any activity to be a successful and a sustainable project. That is why many challenges can be overcome by one solution, as to empower women.

Acknowledgement

Apart from my efforts, the success of this paper depended largely on the encouragement and guidelines from many others. I take this opportunity to express my gratitude to the women and clan leaders from the remotest areas in the Sudan. I went to find the ways and means to help them, and in return they have given their heart. I ended up with a treasure of knowledge. This knowledge and civilization that I would have never acquired in any school or university in the world.

I acknowledge their effort, time and hospitality extended to me, as this is the only thing that I can do. I express my deepest gratitude to them.

I would also like to thank Mr. ElSheikh Amin ElSheikh, Treasurer of the Businessmen and Employers Federation and the Chairman of the UN Global Compact Local Network Sudan, for his support and belief in the work.

I would like to gratefully acknowledge the enthusiastic discussions, of how our efforts can make the change, to all the professionals that had contributed to it in their diverse fields of academics, health and nutrition, law, clergies, civil society (local and international), UN Agencies, administration, anthropologists, historians, government authorities and the business sectors.

My gratitude is also extended to Ms. Asia Makawi Ahmed, the chief librarian at the Ahfad University of Women for her time and assistance.

I could not forget to thank my neighbour and friend, Dr. Eimen Diab, who motivated and encouraged me to write this paper as a result of my ongoing research. I would also thank her for introducing me to the NAM S&T Centre.

Here I wish to avail this opportunity to express a sense of gratitude to The Centre for Science and Technology of Non-Aligned and Other Developing Countries (NAM S&T Centre) for giving us this opportunity of learning and interacting with a widely diverse network of persons who work devotedly to advance our nations.

Finally, an honourable mention goes to my mother, Widad Mohamed Tawfig Hussein, who supported my work, with her great open heart to host our formal and informal discussions. She definitely opened her heart and house for the great cause.

Abbreviations

ICT:	Information and Communication Technology
CSR:	Corporate Social Responsibility
UN Agencies:	United Nations Agencies
CBO:	Community Based Organization
NGO:	Non Governmental Organization
INGO:	International Non Governmental Organization
FGM:	Female Genital Mutilation
FGC:	Female Genital Cutting
HIV/AIDS:	Human Immunodeficiency Virus Infection/Acquired Immuno-deficiency Syndrome

References

1. The National Strategy for Reproductive Health, The Federal Ministry of Health, Republic of Sudan 2006 ~2010, August 2006.
2. Empowering Women for Sustainable Development, the United Nations Economic Commission for Europe, Geneva, Switzerland, By Lisa Warth and MalinkaKoparanova, Jan 2012.
3. Reproductive Health/GBV KAP Survey among Communities Affected by Conflict in Darfur.
4. Author: Dr.EzzeldinRahama, Publication date: 2009, Sponsored by UNFPA..
5. Maternal and Neonatal Health in Sudan: Results of a Situation Analysis Study, Author: Population Council – Sudan, No. of pages: 15, Publication date: 2010.
6. Southern Sudan Maternal, Neonatal and Reproductive Health Strategy.
7. Reproductive Health Officer, Duol Tut, with woman with twins, Malakal Hospital.
8. Government of Southern Sudan Ministry of Health.
9. Action Plan 2008 – 2011.
10. El-Sanosi, Maha. "The Violation of Women's Rights in Sudan: In the Name of the Law?".
11. Think Africa Press. 5 Apr 2012. http://thinkafricapress.com/sudan/violence-against-women-and-sudans-article-152.
12. Humanitarian Response Situation Update.
13. http://www.unfpa.org/emergencies/sudan/
14. Cooking Fuel Saves Lives: A Holistic Approach to Cooking in Humanitarian Settings http://womensrefugeecommission.org/programmes/firewood.
15. Nigeria: Female Genital Mutilation, Inhumanity Against Women.
16. http://allafrica.com/stories/201210290712.html.

Endnotes

- Although Sudan has launched the Central Bureau of Statistics, there is still insufficiency of data and research. Statistics in Sudan is the major problem, since inconsistent patterns in the most of Governmental bodies. For instance, the number of population can be any figure from 34 million to 38 million, varying with the source of information.
- The reliability of the information is also questionable. We always rely on the first hand own collection of data in our researches.
- More children and elderly have died in our armed conflicts than soldiers.
- There is no data for those who died from indirect effects of conflict *i.e.* shortages of food, water and medicine.

- Women and young men on all sides of the conflict were raped, as soldiers raped their own countrypersons.
- Due to poor literacy levels, low education and lack of opportunities, many women still find themselves at the lower end of the job market.
- The system promotes women's involvement in decision making, however, low literacy levels and lack of general opportunities affect women participation
- The development of women organizations are still premature in the field of needed areas. Women empowerment is still viewed from the narrow angle of inequality.
- Qualified educated women have a greater burden in society to assist others in acting positively towards the wellbeing of others.
- Forced early marriage is reported to be a significant problem in Sudan, although information as regards to prevalence is not available. The national legislation of children protection for Sudan has been introduced in 2010, but it does not include protection against early or forced marriage.

Chapter 11

Myanmar Women and Girls in ICT

Khin Mar Lar Tun

University of Computer Studies, Yangon.
14/20, Than Lwin Street, Bahan Township
Yangon, the Republic of the Union of Myanmar.
E-mail: marlartun@gmail.com

ABSTRACT

Both government and non-government organizations in Myanmar are trying hard for promoting the welfare and advancement of Myanmar women in every perspective, and to enable them to participate fully in its national development regardless of nationality, race or religion. All those organizations agree upon the point that information and communication technology (ICT)is the main factor to promote women affairs. The president of Myanmar Women Entrepreneurs' Association (MWEA) pushes women in their economic to use ICT. To achieve this goal, the human resource development in ICT sector carries out ICT education by government ministries, non-governmental organizations and private sector. The Ministry of Education and the Ministry of Science and Technology (MoST) offer many educational programmes for different levels and age for women and girls. The Myanmar Computer Federations (MCF) as well as local private centres also provide ICT trainings for women to expend their knowledge and education on current technology. This paper focus on ICT education programmes for women provided by the government. Regarding the education of Myanmar women, there is at present no gender gap at any level, even in the developing regions.

Keywords: *Myanmar women, Myanmar ICT education, Women affairs in ICT, Women and ICT in Myanmar, Myanmar women organizations.*

Introduction

Women in Myanmar have a unique status since women make up more than 50 per cent of nation's 60 million populations. Their roles in nation become vital in the State's endeavours to build a developed nation. Government has dischargedthe national policies and programmes to utilize the full strength of women both in urban and rural areas (the Ministry of Foreign Affairs/MOFA, 2013). On July 1996, the government established the Myanmar National Committee for Women Affairs (MNCWA) and the Myanmar National Working Committee for Women Affairs (MNWCWA). It was followed by the formation of State, Division, District and Township (grass-roots) levels of working committees for women's affairs throughout the country. Not only government committee for women but also NGO organizations are now being emerged in all sectors for women such as education, health, economy, political, sport and human right. Some well-known organizations are the Myanmar Women Affairs Federation (MWAF), theMyanmar Women Entrepreneurs' Association (MWEA), the Myanmar Women Sport Federation (MWSF) and the Women's League of Burma (WLB).

In Myanmar, no social inequalities can be discerned between men and women and there is no discrimination based on culture, class or colour, although there are regional differences based on local customs.All associations and organizations make concerted effort to promote the status of women both in urban and rural areas. Those national committees laid down the National Plan of Action(draft updated version of 13 May 2010). Its objective is to deal with the issues regarding the advancement of women and respective sub-committees to implement their activities.They include six areas of concern for the advancement of women: (1) Education, (2) Health, (3) Violence Against Women, (4) Economy (5) Girl Child, and (6)Culture.

From these policies it can be seen that the Myanmar Government is pro-active in its efforts to formulate strategies for the advancement of women. The strategic objectives and the National Plan of Actions are set systematically in order to carry out activities for Myanmar Women's Affairs. In this paper only human resource development in education sector for women will be discussed. Regarding the education of Myanmar women, there is at present no gender gap at any level in the developing regions (JICA, 1999).

Women Education

In the education sector, men and women have the same opportunity in Myanmar. Government ministries, private companies and NGOs try their best to promote the education and training programmes that lead to ongoing acquisition of knowledge and skills for women and girls.

Myanmar formal basic education consists of eleven years: primary 5 years, middle/junior 4 years and higher/senior 2 years. After completing basic schooling (5-4-2), the students must take examinations. The total number of students completing secondary school is 47,820 out of which 46.8 per cent are females. After eleven year of high school, students join the higher institutions or universities for 4 years or 5 years for their bachelor degree. They then could continue their studies up to master and doctor of philosophy degrees.

The schools in Myanmar are basically "feminine settings" in that most of the teachersare women. The statistics from DBE showed the comparison of women and men teachers as: 72.9 per cent in primary, 72.9 per cent in secondary (junior) and 70.5 per cent in secondary (senior). Even in higher institutions and universities, mostof the lectures are female (JICA, 1999).

Apart from the formal education, women are also educated through non-formaleducation. The Ministry of Education has implemented non-formal education projects withthe collaboration of the United Nation agencies. The Department of Technical, Agriculture and Vocational Education provide mobile units in the border areas to educate women residing in these areas.

The education policies for women involve: ensuring universal access to primary education both through formal and non-formal education, and to the completion of primary education by at least 80 per cent of primary school-age children;improving women's access to vocational training to provide quality educations and skills to meet the needs of a changing socio-economic context for improving employment opportunities; promoting life-long education and training for girls and women for a broad range of education and training for ongoing acquisition of knowledge and skills as well as support for child care to enable mothers to continue their schooling (JICA, 1999).

ICT Infrastructure

The supreme body of ICT development in Myanmarhas been headed by the Myanmar Computer Science Development Council (MCSDC) since1996.The council has structuredthree layers of model for Myanmar's ICT development activities, such as policy maker, regulator and implementation. The two layers are implemented by government ministries (the Ministry of Communications, Posts and Telegraphs and e-national Task Force) and the implementation layer involves both government and non-governmental organizations (the Myanmar Posts and Telecommunications - MPT, the Myanmar Teleport and the Myanmar Computer Federation - MCF).

According to the statistics of March2012, (TheinOo, 2012) the teledensity factor is only 6 per cent of the population of 60 million, and the total telephone line is 2903 thousand lines. Mobile phones usage is 2500 thousand with five different systems: D-AMPS – 3 per cent, GSM - 35 per cent, CDMA 800 – 18 per cent, CDMA 450 – 42 per cent, WCMA – 2 per cent. There are only two internet service providers (ISP) with service subscribers of over 50 thousand and 2.5 thousand public access centres with over 500 thousand users. Connection systems are based on dial-up, ISDN, ADSL, iPStar, WiMax, WLL, FTTx.

The ICT usage by women in Myanmar is low in comparison to men, *e.g.* the internet usage isly 28 per cent in females with the average age of 25.8. To improve this situation, the Myanmar ICT development Master Plan (2011-2015) set the short-term and long-term goals.The target is to achieve 30 per cent of teledensity in Fixed and Broadband and 25 per cent in mobile space by 2025.

Myanmar's commitment to ICT development is apparent in the establishment of the Yatanarpon cyber city, the improvements of national backbone and international connectivity, the development of government fibre network and e-government projects, computer universities in regional areas, public access centres in the rural areas, and efforts at localization (TheinOo and MyintMyint Than, 2010). The other areas of improvement are the legal framework and the development of relevant applications for both public and private sectors. A key success factor is to devise a mechanism for effective collaboration and cooperation.

ICT Education

Human resources skill development particularly for ICT education, including IT engineers and IT services sectors is considered as an important key factor to Myanmar's economicgrowth. These sectors also provide equal access to quality, practical education for women and girls.A recent survey conducted by the MCF found that 62 per cent of the people in Yangon see the ICT industry as a promising field for their career opportunities.

According to the National Plan of Action for theAdvancement of Women 2011-2015(draft updated version, 13 May 2010) issued by the Ministry of Social Welfare, Relief and Resettlement through its Department of Social Welfare (DSW), the percentage of women completing formal and non-formal education in maledominated spheres (*i.e.* technology and science) and the percentage of femaleheaded institutions/ departments (*i.e.* universities, colleges, ministerial departments)is to be improved further.These issues have been fulfilled by the effort of government ministries and private sectors. The Ministry of Science and Technology (MOST), the Ministry of Education and private sectors are providing thebasic computerskill trainings to professionals. According to the statistics, female are dominant in IT fields since various IT graduatedstudents from all computer universities within the country is 87 per cent, as shown in Table 11.1.

The Ministry of Education offers ICT graduate programmes through its New CenturyHuman Resource Development (NHRD) Department. Example of such programmes are Bachelor degrees in Computer Science Programme at Dagon University (locatedin lower Myanmar) and Yadanabon University (located in upper Myanmar), a Postgraduate Diploma anda Master of Computer Science programme at Yangon University. NHRD Department has also offeredvarious graduate, diploma, and certificate programmes related to ICT and the number of its graduates have been increasing steadily. Among them, more than 35 per cent is female, thus indicated that women in Myanmar already have equal chance in ICT when compared to men (TheinOo and MyintMyint Than, 2010).

The Ministry of Science and Technology (MOST) has also been carrying out the tasks concerning development of human resources to fulfil the increasing demand of industries. The Ministry should develop courses on new academic fields in order to meet the needs of current situation, to enable qualified students to pursue advanced technologies, and to disseminate technological know-how to the entire nation. TheHRD programme involves training to engineers and IT professionalsthrough the Department of Advanced Science and Technology (DAST) and the Department of Technical and Vocational Education (DTVE). The programmes have established twenty six Computer Universities, twenty eight Technological Universities, three

Table 11.1: Gender Statistics of ICT Human Resource Development from Twenty Six Universities of Computer Studies with Respect to the Degree

Degree	*2007-2008*		*2008-2009*		*2009-2010*		*2010-2011*		*Per cent*	
	M	*F*	*M*	*F*	*M*	*F*	*M*	*F*	*M*	*F*
Ph.D. (IT)	1	68	–	12	3	20	3	29	**5**	**95**
Master of Information Science (M.I.Sc.)	–	10	9	69	15	95	2	38	**11**	**89**
Master of Application Science (M.A.Sc.)					1	9	–	13	**4**	**96**
Master of Computer Science (M.C.Sc.)	6	77	26	264	260	415	31	352	**23**	**77**
Master of Computer Technology (M.C.Tech.)	5	12	17	45	18	94	14	89	**18**	**82**
Postgraduate Diploma in Computer Science (D.C.Sc.), Computer Maintenance (D.C.M), Computer Applications (D.C.A)	68	267	63	217	17	44	11	19	**23**	**77**
Honours in Bachelor of Computer Science (B.C.Sc. (Hons)) and Computer Technology (B.C.Tech. (Hons))	261	1186	151	846	126	646	204	798	**18**	**82**
Bachelor of Computer Science (B.C.Sc.) and Computer Technology (B.C.Tech.)	772	1940	644	1712	722	1554	716	1615	**29**	**71**
Total	1113	3560	910	3165	1162	2877	981	2953	–	–
Per cent	**24**	**76**	**22**	**78**	**29**	**71**	**25**	**75**	**16**	**87**

government technical colleges, eleven technical institutes, thirty six technical high school and four technical high schools in twenty four development zonesaround the country.MOST has provided higher institute or university education as well as vocational trainings in engineering and IT fields for secondary level students who could not join highereducation.

Women in ICT Sector

All universities under MOST, located across regions of Myanmar usually use iPStar connection for internet usage. With this low-cost structure of connection, universities in both rural and urban regions can connect each other and share the education resources of lectures, e-books and internet access.

The chance to entrance the university, colleges or vocational training in Myanmar has no gender discrimination, since University entrance is decided by students' performance in high schools examination marks, and students can choose their interests when they have higher marks. Girls in Myanmar usually achieve higher marks in the exam as compared to boys, so that they could choose their fields of interest such as medical, engineering or ICT in the university entrance. Many girls enrol for computer studies in computer universities. Since the universities are established across the country, both students in the rural and urban regions have equal chance to study computer science. According to the statistics shown in Table 11.1, in ICT engineering field, girls' populations is more than boys since 87 per cent of ICT graduated students are girls. Therefore girls in both rural and urban regions are given an equa opportunity for education in ICT development.

The private sector and NGO such as the Myanmar Computer Federations (MCF) and local computer training centres promote humanresource development in the ICT sector. Currently there are more than 100 computerschools, including 80 in Yangon. The National Computing Centre (NCC, www.nccedu.com) also offers InternationalDiplomas in Computer Studies, International Advanced Diplomas, and Bachelors in Computers and Information (TheinOo and MyintMyint Than, 2010).

A survey conducted by the MCF stated that more than 50,000 students study the basic computer skills training and nearly 1,200 students join for international graduate programmes on ICT major annually. Among them, more than 30 per cent is females, thus the women in Yangon notice that ICT industry is a promising field for their future career.

Conclusions

According to the National Plan of Action for the Advancement of Women, women require not only education of ICT fields but also other important matters such as health, economic, emergencies, violence against women, decision making, human rights, and environment. Girl-child areas have been stated and strategic objectives and goals issued for the protection, promotion, and realization of the rights of women and girls as an expression of the Government of the Union of Myanmar's commitment.

Women in Myanmar are also participating in leading business sector of ICT because the Myanmar Women Entrepreneurs' Association (MWEA) encourages women in business and entrepreneurship. MWEA encourages women in business to utilize ICT for communication, sharing and expanding their knowledge and business. Likewise, the Myanmar National Committee for Women's Affairs (MNCWA), the Myanmar Women Sport Federation (MWSF), the Myanmar Women Affairs Federation (MWAF), and the Women's League of Burma (WLB), also actively perform their respective roles in promoting the all-round development of Myanmar women.

Myanmar is doing its utmost to achieve the objectives of the national plan of action for women empowerment at all levelsin social, political and economic development.Myanmar being a developing country with its own struggles, still has a weakness in ICT infrastructure and ICT development for women and girls. Thus, a concerted effort has to be promoted for application of ICT for women both in urban and rural areas. It is ensured that in the near future, the collaborative efforts of all stakeholders could sustain long term success of women's use of ICT and ICT-enabled empowerment and its contribution to their development.

References

1. Aye AyeSoe, 2008. Advancement of Women, Beijing Declaration and Platform for Action, October 2008.
2. ChawKhinKhin, 2007. Myanmar Women Entrepreneurs' Association (MWEA), Consultative Meeting for the Establishment of a Regional Knowledge Network of Rural Women Cooperatives, 23-24 August 2007, Bangkok.
3. Dynamics of Women in ICT Sector in the ASEAN member states, country presentation, September 2011, Jakarta, Indonesia.
4. Japan International Cooperation Agency, 1999.Country WID Profile of Myanmar, Planning Department, JICA, December 1999.
5. Myanmar Women's Affairs Federation (MWAF), http://en.wikipedia.org/wiki/Myanmar_Women per cent 27s_Affairs_Federation
6. Myanmar women's national football team, http://en.wikipedia.org/wiki/Myanmar_women per cent 27s_national_football_team
7. Ministry of Foreign Affairs (MOFA), Women's Affairs in Myanmar. http://www.mofa.gov.mm/aboutmyanmar/wam.html
8. National Plan of Action for the Advancement of Women 2011-2015, draft updated version of 13 May 2010.
9. Status of Myanmar Women, April 2013.http://mncwa.tripod.com/
10. TheinOo and MyintMyint Than, 2010. Myanmar,.mm, Digital review of Asia Pacific 2009-2010
11. TheinOo, July 2012. Japan-Myanmar cooperation opportunities in ICT sector in Changing Myanmar.
12. Women's League of Burma (WLB),http://womenofburma.org/aboutus/

Chapter 12

Empowering Women Skills in Combating Desertification in Sudan

Maha Ali Abdel Latif[1] *and Eiman Elrashid Diab*[2]

[1]*Assistant Professor Researcher,*
Desertfication Research Institute, National Centre for Research,
Khartoum, Sudan
[2]*Associate Professor Researcher*
Environment and Natural Resources Research Institute,
National Centre for Research, Khartoum, Sudan
E-mail: [1]*mahaaali@hotmail.com;* [2]*eimandiab@hotmail.com*

ABSTRACT

The United Nations Convention to Combat Desertification (UNCCD, 1994) addresses the importance of a bottom-up participatory approach in desertification issues, such as in implementing projects to combat desertification. Sudan suffers desertification problem which become much serious after secession of its southern part to an independent country. In decertified areas, the role of Sudanese women and their knowledge in natural resources management and food security become very crucial. They are usually ranked among poorest of the poor and thus suffer problems to sustain their families. Women are affected by desertification directly as it reduce soil fertility and affect water resources upon which they depend to provide food and fuel for their families.

A case study is carried out at the El Rawakeeb Dry land Area with latitudes of 15°-2' and 15°-36' North and longitudes of 32°-0' and 32°-10' East. The system of land use is mainly pastoral except in the low lands where traditional agriculture is practiced. The Ethnic groups of the area belong to Gamuia and Hawaweer tribes. The study is aimed to empower women skills in order to combat desertification through extension activities on desertification cause and effects, and informal seminars on some technologies applied to control desertification, such as: 1. nursing of seedling, 2. planting of trees adapted to xeric conditions and can be used to construct shelterbelts, 3. distribution of *Azadrichta indica* and *Prosopis chilensis* seedlings among women to plant at their homes and giving incentive prizes for best growing trees, 4. provision of gas cylinders for women as energy source to minimize trees cutting. The results show that environmental awareness is improved significantly among women. They participate successfully in restoring plant cover of the area, use alternative energy source safely and willingly share their indigenous knowledge with researchers.

***Keywords**: Desertification, Rural extension, Indigenous knowledge, Rural women, Incentive.*

Introduction

The term "desertification" was initially used by Aubreville (1949). He used it to describe a general process of degradation started with deforestation but not necessarily in dry lands and ending in land turned into desert.

Desertification is defined by the United Nation Convention to Combat (UNCCD, 1994) as a process of land degradation in arid, semi-arid and dry sub-humid areas resulting from various factors, including climatic variations and human activities. Renolds (2001) stated that desertification consists of three major components: meteorological, ecological and social dimensions. The meteorological component relates to drought, atmospheric CO_2, variability of precipitation, and temperature. The ecological dimension includes plant growth, nutrient cycling, regeneration and mortality, plant cover, microbial dynamic and evapotranspiration. Moreover, the socio economic component relates to the loss of habitat, fragmentation of crucial habitat and overexploitation of natural resources *e.g.* over grazing. Desertification induces certain changes as reported by Prince (2002). These changes include: reduced net primary production, high temporal variability and land cover modifications.

Desertification in Sudan

Lamprey (1975) claimed that desertification in Sudan induced by human misuse of land in addition to climate effect. Several studies indicated that about 120 million hectares (ha) of land, including 64 million ha of soils, are desertified to varying degrees. The most desertified zones are the arid and semi-arid zones where 76 per cent of the human population of Sudan lives. Desertification results in destruction of the natural resources and consequently deterioration of the food resources, fuel and pasture.

In desertified areas, women are responsible for all household tasks like water provision, fuel collection, cooking, houses cleaning and cloth washing. Besides, women are responsible to child bearing and rearing (Everts, 1997). Moreover, women are also responsible for grazing animals, seedling and harvesting. These roles are intensified when men migrate to the cities, (Gray *et al.*, 1996). However, despite these responsibilities, some women especially the ones with low income and resources can hardly have access to modern technology, in which may help them to save time and effort in working.

Theories for empowering women show different concepts for women empowerment approaches. The basic approach comes from the Women in Development (WID), in which concern about women problem in terms of their biological differences from men. After some decades the focus has changed from equity to efficiency, and this is based on assumption that: if women involved into productive activities and provided with access to related facilities such as credits and employment, they will be integrated into development process (Tinker, 1992). Accordingly, efforts should be directed towards designing measures to help women in development process mainly by empowering them.

Empowerment is considered as the right to influence direction of change through control of material and non-material resources (Gray *et al.,* 1996). Within these approaches technology as one of the main resources to promote women empowerment has taken different roles, that range from a tool to help women in their work to a tool that enhance women skills and resources approaches.

The present study is carried out at the EL Rawakeeb Dry land Research Station (RDLRS) with the objectives include: empowering women skills to combat desertification through extension activities on desertification cause and effects, informal seminars on some technologies applied to control desertification such as: 1. Nursing of seedlings, 2. Planting of trees adapted to xeric conditions and can be used to construct shelterbelts, 3. Distribution of *Azadrichta indica* and *Prosopis chilensis* to be planted at homes and giving incentive prizes for best growing trees, 4. Provision of gas cylinders for women as energy source to minimize trees cutting.

Materials and Methods

Study Area

The El-Rawakeeb dry land Research Station (RDLRS) has established at El Rawakeeb dry area, in which occupies the area of southwestern Omdurman Governorate. It lies about 45 km away from the capital Khartoum between latitudes 15°-2' and 15°-36' North and longitudes 32°-0' and 32°-10' East.

Climate

According to Abdelatif (2003), El-Rawakeeb area lies in the tropical semi-arid region where the climate is characterized by a short rainy season (July- October) and high evaporation potential. Air temperature values fluctuate and show marked rise in May and drop in July and August due to the incidence of rains.

Vegetation

El-Rawakeeb area lies in semi-desert scrub and the Grassland on Basement complex soils and part of the Acacia Desert Scrub, where can be included in the "*A cacia tortilis – Maerua crassifolia*" subdivision of semi-desert.

Land Use

The system of land use in El-Rawakeeb is mainly pastoral. Traditional agriculture activities are usually carried out, while fodder crop, vegetables and shelterbelts are cultivated and irrigated artificially.

Aspects of Desertification at RDLRS

Desertification gives result in sand dunes formation and sand creep, that cover buildings in the area as shown in Figures 12.1 and 12.2. Desertification also affects plant cover in the area becomes scattered.

Figures 12.1 and 12.2: Sand Covers Houses at RDLRS.
***Source*: DRI, 2012.**

Combating Desertification at RDLRS

Women at El Rawakeeb Area

Women always stay in the El Rawakeeb area because men should go to nearby towns to seek jobs. 97 per cent of the El Rawakeeb women are housewives and illiterate. They take the following tasks to sustain family life:

- ☆ Provision of water for domestic use
- ☆ Collection of fuel wood for cooking
- ☆ Take care for children

Reasons for Choosing Women

- ☆ Some of the activities carried by women to sustain family have adverse environmental such as cutting wood from trees and grazing animals, which intensify desertification process.
- ☆ Women are the ones who face desertification and suffer directly and thus are the ones who accept the idea and can participate directly.

Methods Applied to Empower Women Skills to Combat Desertification

a. Awareness Seminars

The seminars are done through direct communication with women by gathering them and tell them about the causes of desertification, as well as the methods needed to mitigate desertification process. Figure 12.3 shows one of the gatherings of women for awareness.

Figure 12.3: Gathering Women for Verbal awareness.
***Source*: DRI, 2011.**

b. Incentive Method

This involves distribution of *Azadirichta indica* and *Prospis chilensis* seedlings for home planting, and then an annual cash prize is given as an incentive for the best grown seedlings.

c. Technical Training

The women are trained to restore plant cover by planting seeds suitable to be grown in desert environment. They are also trained to prepare seedlings in nursery, as follows:

1. Seeds of *Acacia mellifera*, *Acacia sayal*, *Acacia nubica*, *Acacia tortilis*, *Eucalyptus* sp. and *Conocarpus iancifolius* are planted at the RDLRS nursery.
2. The seedlings are transferred after hardening, to be planted as shelterbelts where seedlings of Acacia species are planted in three rows, followed by two rows of *Eucalyptus* sp. and *Conocarpus iancifolius* seedlings.
3. Gas cylinders with cookers are distributed as the alternative fuel source. Gas cylinders and cookers are provided by DRI in collaboration with the National Authority of Forestry.

Results and Discussion

1. Awareness Seminars

The knowledge about desertification cause and effect is greatly improved when the oral seminars were conducted. Women are persuaded not to cut trees and also to participate in combating programmes. They have showed appreciated efforts to help researchers in planting trees and preventing animal random grazing. Women also share their indigenous knowledge with the researchers in desert control issues. Asgedom (2007) indicated that many opportunities are created to increase awareness of the land degradation and empower women, in order to combat desertification. However, the success of these activities varies within communities and depends on the objectives, ambition, determination and interest, as well as the relation they possess with the community members in the society.

Rossi and Lambrou (2008) illustrated that women know the best trees for fuel, which plants have medicinal uses, where to find water in the dry season, and the best conditions for growing local crops. Thus, they are actually the invisible managers and practitioners in combating desertification.

2. Incentive Outcome

The seedlings grown by women have received great attention and care. This attention is manifested in annual prize given to more than one woman every year. Acronym is usually done at the area to celebrate best growing tree or trees. This celebration is held in synchronization with the World Desertification Day. Figure 12.4 shows one of the best growing tree or trees celebration at RDLRS.

Figure 12.4: The Best Growing Tree or Trees Celebration at RDLRS.

3. Restoring Plant Cover

Women have participated successfully in growing seeds in the nursery and transferred the seedlings to construct the shelterbelts, as appeared in Figures 12.5 and 12.6 respectively. Restoring plant cover induces positive environmental effects,

Figure 12.5: Growing Seeds at RDLRS Nursery.

Figure 12.6: Construction of Shelterbelts at RDLRS.

where sand creep is significantly reduced. The results of Abdel Halim (2003) indicated that the construction of shelterbelts revealed that wind speed was reduced significantly and thus made the sand movement decrease.

4. Distribution of Gas Cylinders as Alternative Energy Source

The action of cutting trees and shrubs in RDLRS by women is greatly influenced the plant cover in the area. According to Abd Allah (2005), most of the women at RDLRS use wood for cooking and very few use it for lighting, whereas the majority of them use wood for both cooking and lighting as given in Figure 12.7.

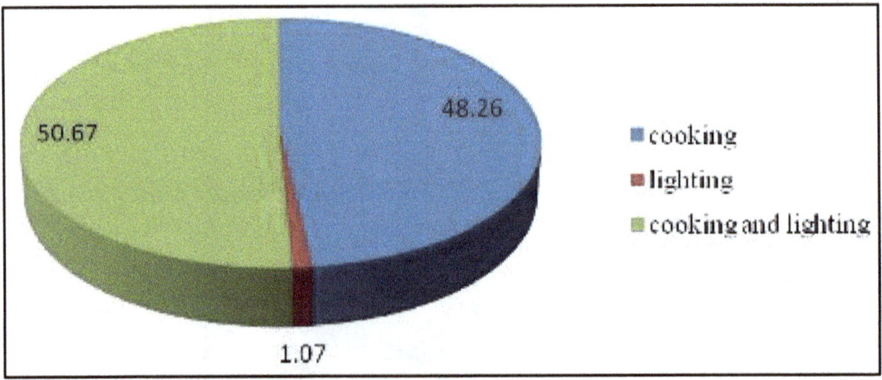

Figure 12.7: Domestic Use of Gas at RDLRS.

EL Hassan (2003) recorded the changes in the number of plant species for two different periods of 1985 and 2000. The data obtained is shown in Figure 12.8, indicates the decreased number of plant species for both tress and grasses. Trees were cut as fuel wood whereas grasses were used as pasture plants.

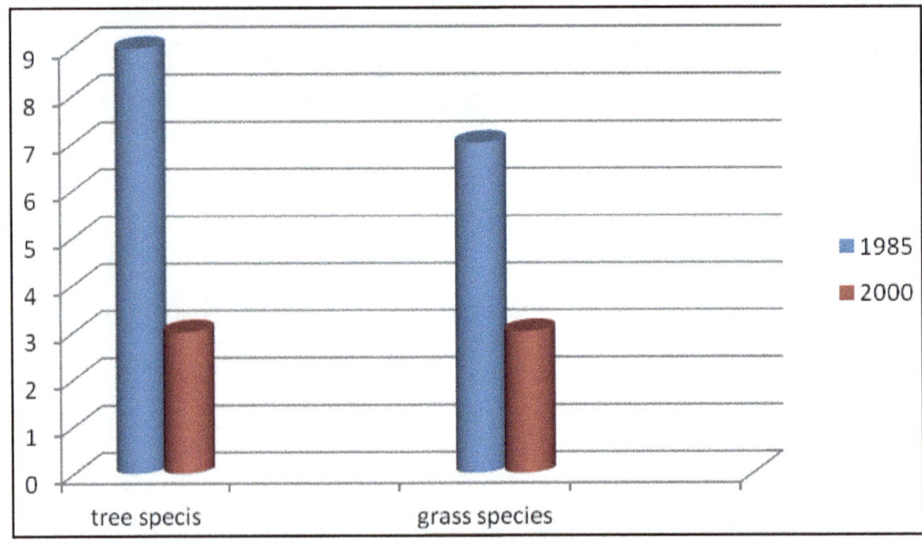

Figure 12.8: Change in Tree and Grass Species as Recorded for the period of 1985 and 2000.

The data obtained by El Hassan (2003) indicated that 100 per cent of women use gas cookers and 27 per cent of women replace them with potages, as gas cookers are easily to be used and saving their time for collecting fuel wood. They have no effect on health as they do not produce smoke and reduce tree cuttings efficiently. The results shown in Figure 12.9 illustrate the time spent to collect fuel wood. It is clear that the majority of women spend twelve hours in collecting fuel wood. By using gas will help women to save their time and efforts to collect fuel wood. The use of gas for fuel has benefited the country in improving health, environment, economy and energy conservation.

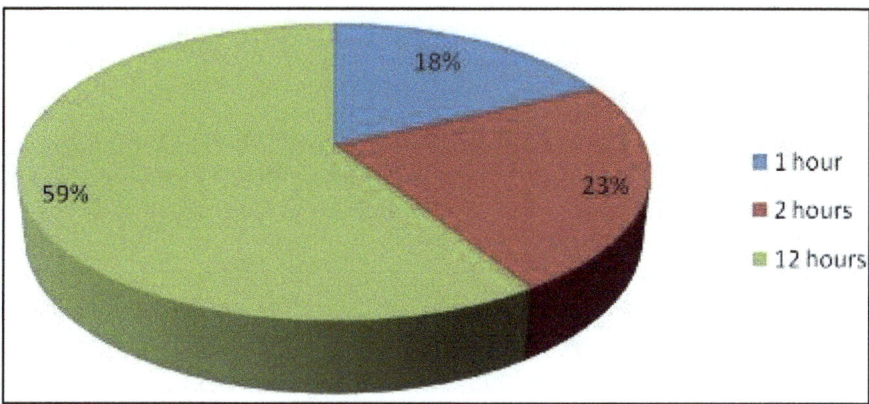

Figure 12.9: The different Time to be Consumed to Collect Fuel Wood at RDLRS.

Conclusions

Desertification is a complex and serious environmental phenomenon. Dry land communities have developed important strategies and repository of knowledge and expertise, which allows them to respond and to survive in challenging conditions. Although women's social position is often subordinate, they perform many essential survival tasks and have developed valuable skills and practices that complement men's knowledge. Severe environmental degradation, however, puts extra burdens on women, who are often left behind to run households when their men migrate to seek jobs. Although traditions and social norms may hinder women's roles in participation and decision making in dry land management, there are many examples where women have organized themselves to combat desertification. These actions include participation in combating desertification through participating in restoring plant cover.

Desertification is a serious problem to be featured by Sudan, that resulting from the interaction of natural and human factors. The problem affects women in different productive roles. While stress and hardship rise for everyone as the nearby natural resource began to deteriorate, women usually spend great efforts to sustain their family. They are the primary custodians of indigenous knowledge system. They have acquired extensive understanding of the natural environment, of its flora, fauna and ecological processes.

The case study carried out at the RDLRS indicated positive roles of women in combating desertification. They participated significantly in restoring plant cover and preventing trees cutting by using an alternative energy source.

Abbreviations

DRI: Desertification Research Institute

RDLRS: Rawakeeb Dry Land Research Station

UNCCD: United Nations Convention to Combat Desertification

WID: Women in Development

FAO: Food and Agriculture Organization of the United Nations

Acknowledgements

We would like to express our deep appreciation to DRI/NCR, for financing this study. Our thanks also extended to the community at RDLRS for keen support and co-operation especially women who shared their knowledge and effort.

References

1. Abd Allah, E. A., 2005. Evaluation of Gas Use in Energy Conservation: A Case Study at EL Rawakeeb Dry Land Area. Published by the EL Neelin University.
2. Abdel Halim, R.E., 2003. Evaluation of Role of Agro-forestry in Combating Desertification. DRI Annual Scientific Report.
3. Abdelatif, M.A., 2003. Impact of Agricultural Practices on Some Selected Soil Decomposers Faun. Ph.D. Thesis, Faculty of Science University of Khartoum.
4. Asgedom, A., 2007. Combating Desertification in Tigray, Ethiopia: Field Study on the Implementation of the UNCCD in the Rural Region of Tigray. Published by Tema vatten i natur och samhälle, Ethiopia.
5. Aubreville, A., 1949. Climates, Forests et Desertification de l'Afrique tropicale, Societe d' Editions Geographiques, Maritimes et Coloniales, Paris.
6. El Hassan, S. E., 2003. Assessment and Adaptation Mechanisms of Rural Economy: A Case Study at EL Rawakeeb Dry Land Area. MSc. Thesis submitted to the Khartoum University.
7. Everts, S. and R. Schulte, 1997. Vietnam Women's Union Promotes Solar Energy. *Energia News* Issue no. 3.
8. DRI Annual Scientific Reports, 2008-1012. Published by the National Centre for Research.
9. Lamprey, H.F., 1975. Report on the Desert Encroachment Reconnaissance in Northern Sudan. 21 Oct. to 10 Nov. 1975. UNESCO/UNEP, Paris/Nairobi. Republished in Desertification Control Bulletin: (17), pp. 1–7.
10. Prince, S.D., 2002. Spatial and Temporal Scales for Detection of Desertification. Published by the Cambridge University Press.

11. Reynolds, J.F., 2001. Desertification, Encyclopedia of Biodiversity. Academic Press, San Diego, pp. 61–78.

12. Tinker, I., 1992. The Political Context of Rural Energy Programmes. Energy for Rural Development. M.R Bhagavan and S Karekezi (eds) Zeb Books London, UK.

13. UNCCD (United Nations Convention to Combat Desertication), 1994. United Nations Convention to Combat Desertication in Those Countries Expressing Serious Drought and / or Desertification, Particularly in Africa. Secretariat of the UNCCD, Bonn, Germany, p. 76.

14. Rossi, A. and Lambrou, Y., 2008. Gender, Equity and Social Issues in Liquid Biofuel Production: Minimizing the Risk to Maximize the Opportunities. Published by FAO, Italy.

APPENDIX
Species of Plants Recorded in Soil Seed Bank at EL Rawakeeb Dry Land Area

Sl.No.	*Botanical Name (Local Name)*	*Family Name*	*Life Form*
1.	*Aristida adscensionis* (El Gaw)	Poaceae	Annual
2.	*A. hordeacea* (Gaw)	Poaceae	Annual
3.	*Sporobolus pyramidatus* (Aish el far)	Poaceae	Annual
4.	*Tragus beteronianus* (Um gandowl)	Poaceae	Annual
5.	*Dactyloctenium aegyptium* (Um assabi)	Poaceae	Annual
6.	*Enneapogon brachystachyus* (Adom Gash)	Poaceae	Perennial
7.	*Cenchrus biflorus* (Haskaneet)	Poaceae	Annual
8.	*Eragrostis megstachya* (Gadafa)	Poaceae	Annual
9.	*Echinochloa colona* (Difera)	Poaceae	Annual
10.	*Indigofera hochstetteri* (Sharaia)	*Fabaceae*	Annual
11.	*Crotalaria saltiana* (Sofaira kabira)	*Fabaceae*	Perennial
12.	*Amaranthus graecizans* (Lisan tair sagair)	*Amaranthaceae*	Annual
13.	*Corchorus tridens* (Khudra)	*Tiliaceae*	Annual
14.	*Gisekia pharnacioides* (Safal)	*Phytolaccaceae*	Annual
15.	*Glinus lotoides* (Turba)	*Molluginaceae*	Annual
16.	*Zaleya pentandra* (Rubaa)	*Aizoaceae*	Annual
17.	*Ipomoea verticillata* (Tabar)	*Convolvulaceae*	Annual
18.	*Citrullus colocynthis* (Handal)	*Cucurbitaceae*	Annual

Source: DRI Annual Scientific Report (2012).

Part III

Best Practices and Case Studies on Gender Equality and ICT

Chapter 13

Country Status Report on ICT Development in Gambia

Bintou Dibba

Principal Science, Technology and Innovation Officer, Ministry of Higher Education, Research, Science and Technology, Kotu, The Gambia
E-mail: binsdd@gmail.com

ABSTRACT

Information and Communication Technology (ICT) has been one of the fastest growing sectors of the Gambian economy, as it contributes to GDP growth and poverty reduction. It has been the goal of the Government of The Gambia in its vision 2020, to transform The Gambia into technologically advanced and information rich society.The nation's Information and Communication Infrastructure (NICI) policy and plan, the ICT4 Development (ICT4D) Action Plan, the e-government strategy and the enactment of the Information and Communication Act 2009,provide a strong and conducive foundation to an ICT-led environment that is beneficial to the socio-economic wellbeing of the country and its people.

The Gambia in the last 10 years has provided the right path to realizing government's Vision: *A Gambia with the requisite infrastructure and enabling policy framework that ensures full connectivity of everyone to ICT services.* The goal is to turn The Gambia into a competitive ICT incubator and hub, which would become an outsourcing destination and a base for technology-based services, knowledge production, transfer and development.

The ICT sector in The Gambia spans over areas such as telecommunications, internet, information technology and television/radio. In the telecommunications sub-sector, telephone penetration level (tele-density) has shown steady growth in the last three years. The Gambia telecommunication sub-sector has one fixed line operator and four GSM service providers. In addition, the number of registered internet service providers has risen to six.

The media sub-sector has also registered significant progress both in terms of number and coverage. The Gambia Radio and Television Services (GRTS) went on satellite in 2009.

The number of FM commercial and community radio stations as of June 2011 was 19 and 9 respectively. Also the number of print and electronic media has increased significantly. The progress made in this sub-sector means that more women especially women in the rural areas can now have access to information in agriculture, health, education and other sectors that will improve their socioeconomic wellbeing.

Despite the progress that has been made, further developments in the sectors are facing some major challenges, such as: expensive and inadequate internet connectivity; difficulty in extending communication and internet services to other regions of the country due to obsolete national telecommunication infrastructure and less access to nationwide radio and TV coverage. The country is still struggling in its efforts to adopt and implement the e-government programme that was initiated in 2007 through the support of UNECA.

Keywords: *Telecommunications, Media, Information, Gender, Women, ICT.*

Introduction: Demography

The Gambia is a small country in West Africa, surrounded on three sides by Senegal and bordered on the west by the Atlantic Ocean. According to the Population and Housing Census, The Gambia's projected population is estimated to be 1.79 million (2011 estimation), with an annual growth rate of 2.74 per cent. The population constitutes 50.7 per cent female and 49.3 per cent male. This population also demonstrated 42 per cent of the population is youth below 15 years of age, and 22 per cent between 15 and 24. Only 3.4 per cent of the population is 65 and over (2011 estimation). The country has a population density of 133.63 people per square km, ranking number 74 out of 230 countries.

Status of ICT in the Country

Recent trend in the country's GDP growth shows ICT sector as one of the fastest growing sectors of The Gambian economy. Its contribution to poverty reduction has been impressive with a total contribution to GDP of about 7 per cent.The Government is aware of the fact that accessibility of ICT to the population at large and effective ICT usage is imperative to the advancement of the current programmes. It has been the goal of the Government in its vision 2020 to transform Gambia into a technologically advanced and information rich society.

The Gambia Government enacted the National Policy for the Advancement of The Gambian Women (NPAGW) in 1999. The policy provides a legitimate point of reference for addressing gender inequalities at all levels of government and by all stakeholders. Through the continuous usage of ICTs such as TV, radio, print media, as well as computers and the internet in schools, major achievements were registered in the areas of policy concerns and pronouncements. These have increased awareness on gender as a development concern; increased enrolment and retention of girls in schools; improved health care delivery; increased women participation in decision making, and reduced gender stereotyping and discrimination.

Studies have shown on an average, the overall ICT access within the population has increased. Due to the expansion in this sector and the work of IT associations, the proportion of students and teachers using internet services is very high. About 98.4

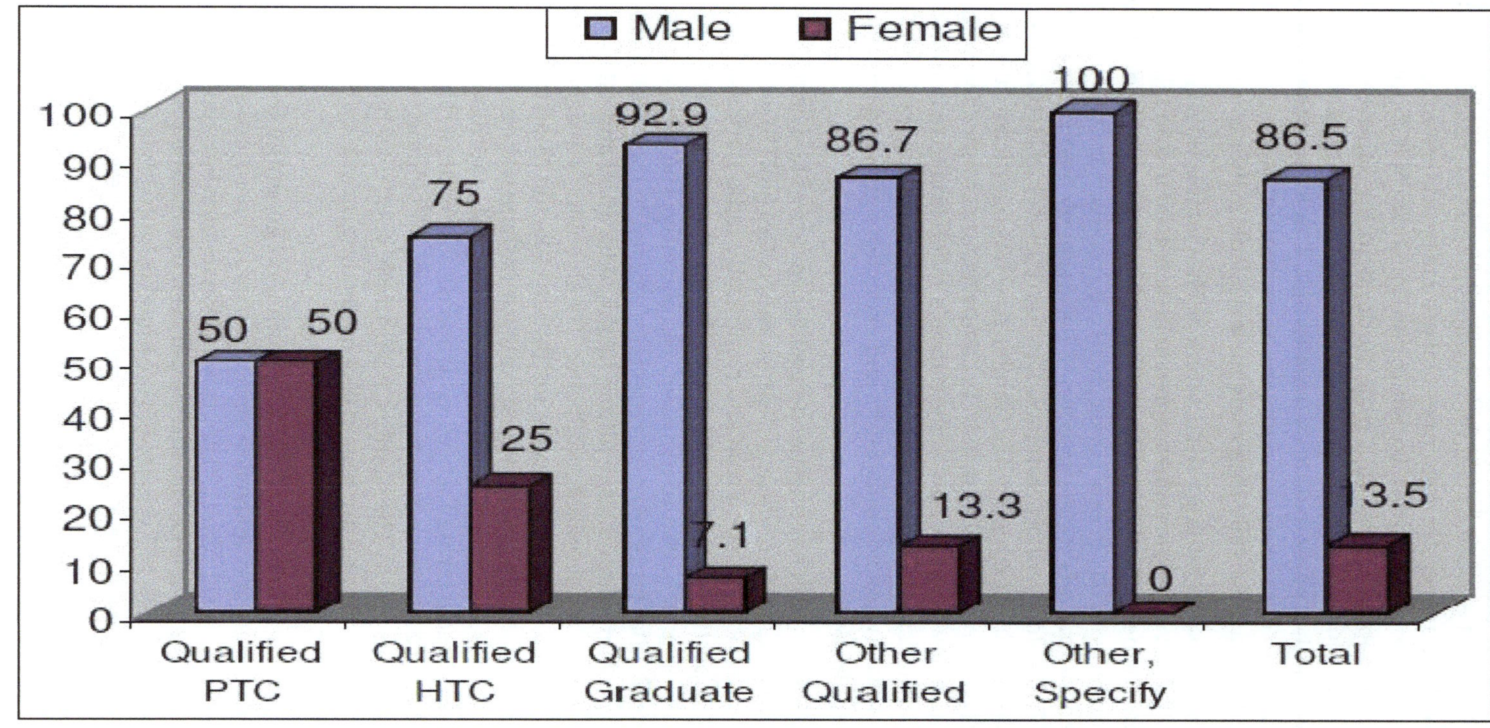

Figure 13.1: ICT–Qualified Teachers by Category of Qualification and by Gender.

***Source:* Scan-ICT Survey 2006.**

per cent of students and 98.9 per cent of teachers used internet services for email messaging (Scan-ICT Survey, 2006). With regards to training received in the use of ICT, however, there are gender disparities that need to be addressed. Only 18 per cent of female students enrolled in tertiary education which are in ICT dominated fields, while there were about 60 per cent male students in the same category, *i.e.* ICT dominated fields. However, the results that emanated from the Vocational/Technical schools show a higher participation of females in ICT field than the male students. There were 17 per cent female students in this school category, which are in ICT dominated fields as against only 15 per cent male students in the same school category.

The Information Technology Association of The Gambia (ITAG), established in 2004, aims at more young girls and women to be involved in and take up careers in ICT. ITAG was formed with the mandate to provide technological know-how of basic ICT applications in everyday lives of young girls. ITAG conducts annual ICT programmes to educate and inform girls about careers in ICT and schools have benefited from their numerous ICT development programmes. Through the work of ITAG and many other IT associations within the country, The Gambia now has significant number of female students involved and specialising in ICTs compared to a couple of years ago. There are more female news casters now at the nation's TV station than before, more girls practising journalism, more girls specialising and working in institutions and organisations as network administrators and even more girls engaged in the sale of credits for mobile users.

The media sub-sector has registered significant progress both in terms of number and coverage. The Gambia Radio and Television Services (GRTS) went on satellite in 2009. The number of FM commercial and community radio stations as of June 2011 was 19 and 9 respectively (PAGE 2012-2015). The number of registered newspapers (both print and on line) have also increased significantly. The progress made in this sub-sector means that more women, especially women in the rural areas, can now have access to information in agriculture, health, education and other sectors that will improve their socioeconomic wellbeing. The development agents such as non-profit organisations, NGOs and public institutions like the Ministry of Health and the Ministry of Agriculture, have been making use of these stations in disseminating important key issues information, such as: weather forecasts, nutrition, climate change, appropriate and timely farm practices, crop and disease control. The information is also intended to reach women in order to improve their livelihoods and that of their communities in general. In The Gambia, particularly rural Gambia, women are the custodians of the family and are responsible for feeding, health care, and household maintenance among other matters.

The households maintain access to information generally through the use of radio, mobile phones and television programmes for increased productivity and growth. At national level, about 90 per cent of households have access to radios, 40 per cent have access to television, and 56 per cent use mobile phones to send and receive information. This ICT equipment are found in both urban and rural areas, but their proportions revealed disparities according to place of residence as shown in the Table 13.1.

Table 13.1: Estimated Percentage of Households in Gambia with ICT Access to ICT

House Hold Access to ICT Facilities	*Per cent of Urban Households*	*Per cent of Rural households*	*Per cent Total population*
Radio	93.05	89.31	89.98
Television	82.13	30.57	39.82
Land phone (Fixed)	37.23	9.26	14.27
Wireless Land phones(Jamano)	5.96	3.85	4.23
GSM Mobile	81.83	50.32	55.97
Computer	7.71	1.06	2.25
Internet	5.01	0.24	1.10
Electricity	68.43	1.75	NA

Source: SCAN-ICT Survey 2006.

Challenges

ICTs in The Gambia may fulfil the needs of most individuals and households in products or services delivered to them, however, the direct application of ICTs is still relatively limited. There is only one public TV station and no private TV station in The Gambia. The high cost of computers and connectivity keep ICT services far beyond the reach of most women in the rural areas. ICT infrastructure is largely urban-centred. In addition, gender roles and family responsibilities have translated to the fact that most women have less time to make use of ICTs. The availability of computers in the education sector is still at an average very low. The other big hindrance to ICT development in the country is the less affordable and reliable source of energy that demands a high investment. The cost of computers and the price of internet access are one of the main issues that inhibit the spread of ICTs within the country. The high prices of equipments are still far beyond the reach of the average citizen. The high price tags compared to other necessities in life, act as a barrier in using the ICT equipment and services.

Women's access to and control of ICTs has featured prominently in both theory and practice of the gender digital divide. Women's access to and control of ICTs is affected by factors such as gender discrimination in jobs and education, social class, illiteracy and geographic location, which means that the great majority of The Gambian women have quite a limited access to ICTs.

Conclusions

Recognizing that "information is power" to empower and power to act, the Government of Gambia realized that it has to promote and strengthen the widespread development and use of ICTs in the country. The government has awakened to the fact that the promotion of ICT is also to empower The Gambian women. Though the proliferation of ICT infrastructure and equipment will provide a future base and is an essential tool in bridging the gap between the developed and developing world, it is important to recognize the required relevant knowledge base. It requires, therefore, competitive skills that in particular are lacking among women in the developing world.

The government of The Gambia has put lots of effort in creating an enabling environment for ICT development, established by the Ministries of Information and Communications Infrastructure and Higher Education, Research, Science and Technology. The efforts include: developing ICT policy; formulating National Science, Technology and Innovations policy and their respective implementation strategies; as well as introducing reforms in the telecommunication sector. However, the dynamics of the national economy requires more substantial changes to satisfy the needs and expectations of the ICT market. To address this, the government is taking appropriate steps in order to accommodate the present and future ICT needs and related initiatives in the country, taking into account the regional and international development trends in the ICT sector.

Abbreviations

FM: Frequency Modulation
GDP: Gross Domestic Product
GRTS: Gambia Radio and Television Services
ICT: Information Communication Technologies
IT: Information Technology
ITAG: Information Technology Association of the Gambia
NGO: Non-Governmental Organisations
NICI: National Information and Communication's Infrastructure
NPAGW: National Policy for the Advancement of Gambian Women
TV: Television
UNECA: United Nations Economic Commission for Africa
GBOS: Gambia Bureau of Statistics

References

1. Republic of The Gambia, Programme for Accelerated Growth and Employment (PAGE) 2012-2015.
2. The Gambia, SCAN-ICT Baseline Survey 2006, September 2007.
3. Ministry of Women's Affairs; Gambia National Gender Policy 2010-2020.
4. Gambia Bureau of Statistic, Population and Housing Census Report 2003.

Chapter 14

Information Technology for Women Empowerment: The Case of *Suvidha Kendras* under Mission Convergence

Rashmi Singh

Executive Director,
National Resource Centre for Women
National Mission for Empowerment of Women,
Ministry of Women and Child Development, Government of India,
Room No. 119, Janpath Hotel, New Delhi – 110 001, India
E-mail: rashmi.singh@nic.in, rashmi.nct@gmail.com

ABSTRACT

Information is power and lack of information is one of the primary factors which keep millions disempowered. Women can be empowered by enhancing their knowledge, skill and access to information and technology. Information Technology (IT) has the potential to reach those women who have not been otherwise benefitted from Government interventions through bridging of the digital divide. Mission Convergence (MC) emerged as an innovative initiative of the Government of Delhi for infusing new governance architecture for actively engaging citizens with a focus on increasing women's participation. IT for masses and not for classes drove the Mission Convergence programme to make investment in human resources, hardware and software on a scale which was unprecedented for any social sector project in Delhi. This paper elucidates how the expanded use of technology in a Public-Private-Partnership (PPP) model enabled the system to include poor and vulnerable women living in slums and other underserved areas, in the ambit of Government's welfare programmes and other public services. The business process reengineering in the programme led to simplified, transparent and user friendly processes which were customized to meet the needs of common citizens. Empowering resource poor women to use and avail IT in the Gender Resource Centre (GRC) - *Suvidha Kendras* helped overcome the socio-cultural barriers hitherto precluding women's equal access to opportunity in the areas such as health, education, skill and social protection provisions. The programme has demonstrated the unbound potential of Information Technology in empowering vulnerable and poor women by addressing their barriers of inclusion, and mainstreaming them in public policy.

Keywords: *Women, Information Technology (IT), Suvidha Kendras, Gender resource centres, Stree-Shakti Kendras, Samajik Suvidha Sangam (SSS), Mission convergence (MC), Delhi.*

Introduction

In the current context, Information Technology (IT) is a very vital tool for inclusive growth. There is no doubt that rapid development of IT has contributed significantly to economic growth and development of the country. The dynamism of IT offers a fundamental change in all aspects of life, including knowledge dissemination, social networking, business practices, political engagement, education, health and entertainment. IT has the potential to reduce poverty and improve livelihoods by empowering users with timely knowledge and appropriate skills (Key, 2000, Torero *et al.,* 2005). IT for "resource poor" signifies a capacity building process that enhances the ability of the community to ascertain various possibilities in livelihood strategies. IT can reduce poverty by improving poor people's access to education, health, government and financial services (Cecchini and Scott, 2003). However, it is seen that access to IT is often with the middle and upper class urban men in comparison with women, who largely tend to remain excluded because of lack of equal opportunity and enabling environment which includes restricted mobility, lower literacy levels, and the patriarchal mind-set which casts traditional gender-stereotypes. Restricted access to communication infrastructure, high access cost, and ignorance about its potential have bestowed the benefits of IT on the better off, urban segments of population and less to the poor, and even lesser for poor women. At the same time the poverty reduction potential of IT requires to be unleashed through sensitive public policy formulation and a careful project design. In the Delhi model, a need was thus felt to ensure that IT infrastructure and its application is inclusive so as to bring in its ambit the poor and needy, especially poor women and girls from the disadvantaged and vulnerable community. Using the potential of IT for poverty reduction by bringing in its ambit of intervention the poor, particularly poor women is a difficult endeavour, but very significant for accelerating the pace of developmental outcomes and achieve women empowerment.

Information Technology for Women Empowerment

Information is power and the lack of information amongst women is one of the primary factors which keep women in a disadvantaged and vulnerable position. IT is for everyone and women have to be an equal beneficiary to the advantages offered by the technology, and the products and processes which emerge from their use.

Women's access to education, health and social services becomes challenging due to the interplay of certain complex socio-economic and structural factors. A creative use of IT can address many of such barriers and create a level playing field with incremental benefits cutting across sectors. For instance, women's involvement in economic activities is largely in informal and unorganized sector employment, including self-employment through farming, crafts production, petty trading, engagement in conventional trades such as cutting, and tailoring etc. which all reflect typical gender stereotyping. There are numerous possibilities for IT to improve women's economic activities through innovative range of options in which women can get engaged in formal as well as informal ways. IT brings lot of opportunities to women in the work situations and handling small business from home with flexi timings that provides the required work-life balance for them. These are some other gender dimensions of IT usages which should be and were encouraged and tried.

Women should be empowered by enhancing their skills, knowledge and access to IT. This will strengthen their ability to combat negative portrayals of women internationally and regionally to challenge instances of abuse of the power of an increasingly important industry (VAPS, 2004). IT includes the use of internet, web-applications, video films, community radio, etc. which have the potential to reach those women who earlier have not been reached out by any other means, thereby paving a way for them to participate in economic and social progress and make informed decisions on issues that affect them. Bridging the digital divide among community with ICT usage becomes a tool for bridging the gap in the knowledge and skill levels of the rich and the poor alike, and for women whose opportunity to avail the full benefits of the knowledge economy becomes limited due to mobility constraints. The benefits of IT usage for women are proportionately higher compared to men. For Government's schemes where the challenge is to reach out to the most needy and underprivileged with standards of transparency and objectivity in the otherwise complex, and non-transparent system, IT holds even greater promise to restore objectivity, and transparency.

IT can work as a novel opportunity to enhance women's access to information and knowledge, as women are the least educated and marginalized especially in the developing countries (Sharma, 2001). Poverty reduction and women empowerment potential of IT requires attentive public policy formulation and a careful project design. Mission Convergence (MC) Programme implemented through a body called *Samajik Suvidha Sangam* (SSS) of Government of Delhi was able to achieve encouraging results in this direction. Innovative institutional mechanisms with successful IT initiatives for women adopted by the MC were characterised by local ownership and participation of the community. Here, IT was used for creating a robust Management Information System. Use of IT overhauled the way various social sector activities were reported and monitored, both online and offline. A dedicated portal for the MC programme helped in online reporting for a gamut of services accessed, availed, and delivered. *Suvidha Kendras* established under the MC used IT effectively and innovatively to make it possible to realize one of the primary objectives of the programme *i.e.* to empower vulnerable and marginalized women. This could be done by demystifying technology amongst thousands of women and girls, giving them hands-on experience, and facilitating their access to government services and programmes which were earlier difficult for them to traverse.

Samajik Suvidha Sangam-Mission Covergence–The Case of Institutional Innovation

Samajik Suvidha Sangam (SSS) - *Mission Convergence* (MC) emerged as an innovative initiative of the Government of Delhi in the backdrop of infusing new governance architecture for actively engaging citizens, with a focus on increasing women's participation. This flagship programme of the Government of NCT of Delhi, is dedicated towards improving the access to public services for the otherwise deprived vulnerable populace of Delhi. It is a convergence model involving GOI-NGO-Community partnership with a focus on creating single window platforms for reaching the development programmes and social sector services to the vulnerable and extremely vulnerable sections. The Mission at the outset generated a database of vulnerable poor through a community based census of people living in vulnerable locations of the city based on a unique set of vulnerability criteria. Besides covering the organized and unorganized slums and the other prima-facie vulnerable locations, a database of homeless of Delhi was also generated through a unique socio-economic survey. This laid the foundation for interventions for such category of people who were largely excluded from the Government's programmes.

The genesis of the programme lay in reorientation of thinking on the very approach of the existent delivery/implementation mechanisms of the social sector

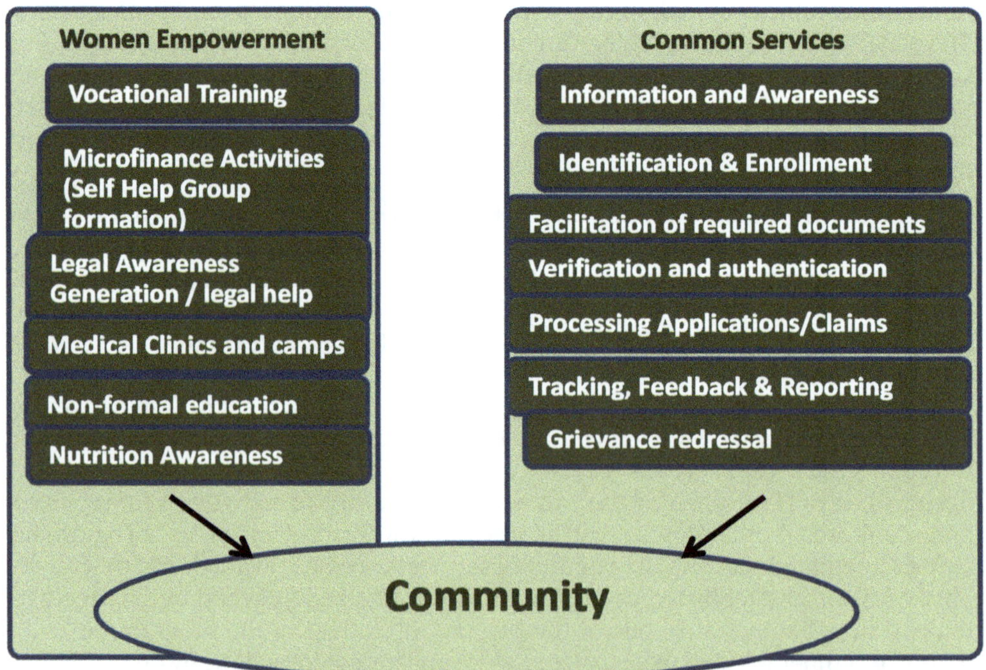

Figure 14.1: Gender Resource and Common Facility Centre.

programmes of the Government with an objective to improve the quality of life of citizens especially the most vulnerable and disadvantaged sections of society. Initial consultations lay stress on the defining the vulnerable people, identifying them through a community based survey, and streamlining the processes and procedures in the actual implementation of welfare schemes of the Government. Thus, the programme envisaged the creation of single window service delivery systems located in the most marginalised communities of the city where the services of nine Government Departments could be synergized and brought on a unified platform. To take this forward, an already existing institutional model of Gender Resource Centres (GRCs) were used as they were decentralized, located within the vulnerable locations, with a proven model of collaboration between the government and NGOs who would run the centres using standards set by the Government, within the overall framework of gender justice and inclusion. However, the number of GRCs had to be scaled up nearly four times. These centres had to be restructured and strengthened with resources when the programme was shifted from Department of Social Welfare/Women and Child Development to *Samajik Suvidha Sangam* (SSS), an autonomous agency of the Government conceived and set up to implement the unique convergence programme in a mission mode.

GRCs became the essential building blocks, and community nodes of the MC programme with an added structure of *Suvidha Kendra* built into the design of the GRC to make these centres serve the community in a more robust manner than envisaged in the original design when the GRCs were conceived. One of the key elements of the design were to include CSOs in the identification of the target group, management of the database of the un-served and the under-served living in the catchment area of a centre, through a Public Private Community Partnership (PPCP) model and work in tandem with the different Government departments who were partners in the Mission. The overriding principal behind the initiative was to create better opportunities for the poor especially the women and girls living in identified locations, improve their access to public services and also create mechanisms for improved service delivery with a system for greater accountability through use of IT and strong community engagement. The process of infusing objectivity in the process of determining the clients to be served, creating unified platforms, involvement of CSOs in Government delivery systems etc. were politically sensitive issues and were negotiated with much care within the programme. Despite this, a lot of debate and discourse got generated around the new architecture. The GRCs were not only provided with resources but also capacity building needs to make them capable of addressing multiple dimensions related to women empowerment in a holistic manner. These dimensions included initiatives towards simultaneously striving for social, economic, and legal empowerment of women particularly those belonging to the under privileged sections of the society.

MC created a network of **125 Gender Resource Centres** - *Suvidha Kendras* (GRC-SKs) including Extension Counters working in the concentrations of vulnerable population. The GRC component includes a set of holistic activities for empowerment of the community women covering interventions in the areas of non-formal education, mainstreaming to regular schools, remedial education for children, health services, nutrition counselling and awareness, vocational trainings, creation of micro enterprise and micro credit opportunities, legal awareness and legal counselling. The *Suvidha Kendras* on the other hand provided information on various schemes of the Government, linkages and facilitation with different services and schemes of the Government through the easy access to information, application forms, facilitation in getting the documentary requirements and certifications as a precursor to completing the forms fulfilled, linking up with the departments concerned through the office of the area Deputy Commissioner, etc. Additionally, a Programme Management Unit (PMU) set up at the central level to anchor the whole programme was also entrusted with the direct implementation of two central schemes namely *Swarna Jayanti Shahari Rozgar Yojana (SJSRY)* and *Rashtriya Swasthya BimaYojna (RSBY)* at an early stage of the Mission.

This State level body called *Samajik Suvidha Sangam (SSS)* was registered under the societies registration act and a Governing Council chaired by chief secretary was constituted as an empowered body comprising of secretaries of the Departments of finance, information technology, planning, administrative reforms, labor, revenue, social welfare, health and family welfare, women and child development, SC/ST/OBC/ Minorities and food and civil supplies. This body also played the role of a mentor and guide to the Mission, while recognizing the evolving nature of the programme and the need to have it continuously innovate with the dictum of learning by doing, instead of expecting the design to be perfect at the outset. The PMU was headed by a Mission Director; a government officer especially selected on the basis of previous experience in social sector programmes, and demonstrated ability of setting up structures and institutions requiring multiple stakeholders' collaboration. In order to develop the programme from conceptualization to the design stage and thereafter carry the baton of implementation, the then chief secretary, Delhi choose an officer who had led the Stree Shakti programme in the years 2004-2007 to a new height and got the international acclaim for winning the CAPAM (Commonwealth Award for Public Administration and Management).

The Mission Director was given the flexibility to choose her team of experts following a process of transparent recruitment from the open market instead of being restricted to work with a team of officials who would be posted in the normal departmental mode as in other departments of the government. Later a handful of government officials were involved through deputation. Norms of engagement were developed for a synergistic working relation between experts and regular government functionaries, based on some initial surface tensions between the two types of human resources in the Mission. The PMU became a vibrant body which was continuously liaisoning with the multiple partners involving the Line ministries, District administration, NGOs, INGOs and other experts under the overall policy guidance given by the Governing Council of SSS and the State Convergence Forum headed by the Chief Secretary, Government of Delhi. This was the apex governing body of the SSS responsible for taking major decisions to facilitate effective implementation of the Mission Convergence Programme. As the programme grew in size and stature, a Policy Review Committee headed by the Chief Minister of Delhi, was constituted. The Chief Minister of Delhi took close interest in the programme, especially when the potential of the programme to make a wide impact on the lives of thousands of women in a positive manner unfolded by each passing day. The Mission Director was made the member secretary of the Governing Council of SSS, and of the policy review committee to develop suitable agendas and follow up on the decisions taken in this empowered body.

The programme was formally launched on 14th August 2008, though a small team at the Delhi secretariat had started working on the prgramme design and setting up the implementation structures since February, 2008. As of now this programme is fully funded by the State government and over 125 *Suvidha Kendras* set up in partnership with credible CSOs, are all equipped with complete IT infrastructure like computers, broadband internet, UPS, printer, and scanners. The selection of CSOs also entailed evaluating the previous experience of the organization in running multi-disciplinary progarmmes which involved multiple stakeholders and effective Management Information Systems. These centres thus became increasingly engaged in the process of resource mapping, identification of the target group in the local community, disseminating information, and creating awareness about public services/schemes besides enrollment and handling grievances.

The institutional mechanism also involved setting up district level centres called District Resource Centres (DRC) located within the DC Office to facilitate, coordinate and monitor the implementation of Mission Convergence programme at the grass-roots level in the respective districts. These centres were equipped to maintain the database of district vulnerable population based on the vulnerability survey undertaken in the respective district, and use this information base to develop targets for the district and the required budgetary resources. From a departmental mode of working, there was a shift to district mode of working for all the welfare/development schemes and programmes. The role of District Collector changed from that of a regulator to a development administrator who had tools to reach out to the vulnerable and ensure their coverage in the government programmes with the help of grass-root level organizations running the GRC-SK's in the respective districts.

The district set up also supervised SSS vulnerability survey process and provided support to the GRC—SKs for facilitation of linkages with existing government programmes spread across different departments. Identification and reaching the vulnerable people through door-to-door survey, facilitation in getting identity documents/ other relevant documents and certificates as required for different schemes became one of the primary mandates of these centres, generating a new found confidence in the poor that the system is reaching out to them. District Convergence Forum meetings started getting organized on monthly basis under the Chairmanship of the respective district collectors to take up issues thrown up by the GRC-SKs who were working at the grassroots level in the respective districts. Handholding, support and monitoring of the activities of the GRC-SKs on the ground were also assigned to few Mother NGOs and experts within the PMU who were designated as district nodal officers for different districts. These mother NGOs were responsible to train, build capacities and submit regular evaluation reports based on identified parameters that would capture qualitative and quantitative progress of various GRC-SKs. Timely submission

Vocational Courses

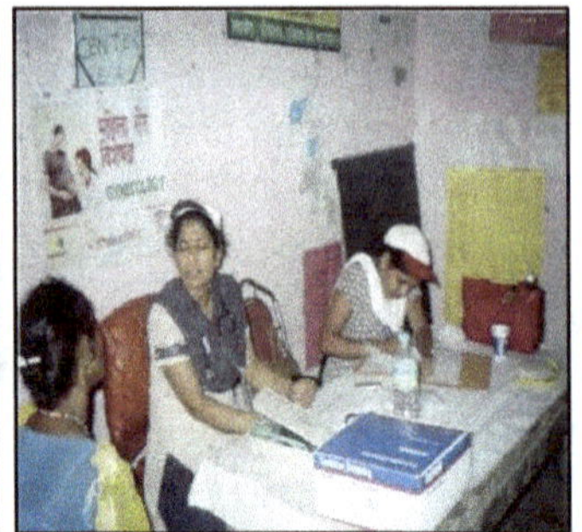

Health Services

Education Classes

Legal Counseling Session

SHG

Information Facilitation

Figure 14.2

of data, and analysis of data was possible only through the adoption of an advanced web enabled MIS, and training of all stakeholders to handle the same.

Mission convergence used the learning's of the *Stree Shakti* programme, a flagship scheme of the government under *Bhagidari* for women empowerment using public-private partnership approach. Under this initiative, the Government in collaboration with non-government organisations (NGOs) held camps for women in various localities of the city so as to empower women from the economically weaker sections through interventions in the areas of health, legal literacy and livelihood promotion. Even though these camps had enrolled millions of women and covered them through various services, one of the gaps in the programme was lack of adequate use of IT to maintain database which could be used for development programmes in a systematic way. The shifting of location of the camps every successive month and later after every 3 months did not provide the scope of making sustained interventions for the development of women. This led to creation of regular centres that enabled to serve the needy pockets in a sustained manner. The experience also gave important insight into the need for adopting a good data management system backed with technology support right from the outset since the entire data of *Stree-Shakti* was being maintained in a manual mode.

While a beginning was already made, the institutional architecture received a major fillip with the initiation of a mission mode programme which at the outset recognized the need for embedding such community nodes focussing on women at the base of the institutional architecture conceived for the programme. The Mission also recognized the need to make investment in technology solutions and systems at the outset itself. In order to have a robust outreach mechanism covering the entire city, MC multiplied the earlier numbers of only 26 GRCs (Gender Resource Centres) to 125, with 20 other extension centres all over Delhi. Poverty maps were created for the city, and the locations for the centres were planned in a manner that the communities for whom such services were required had easy access to the same. This was carefully done through a process of community mapping and use of digital tools to identify areas where *prima facie* concentration of disadvantaged, vulnerable poor households existed.

While the MC programme expanded, it also brought about several improvements in the design and led to augmentation of budget and human resources required for GRCs. The mandate of GRCs was enhanced so that they become a truly robust instrument of community outreach with the addition of *Suvidha Kendras*. In the human resource planning, the role of women as help desk manager, and community outreach worker was carefully built into the design stage and guidelines issued to bring standardization in all the centres (development of Operational Guidelines and SOPs which became the basis for all trainings and capacity building sessions). Women from the community were involved as field surveyors in huge numbers. In order to equip the centres in discharging their expected roles and responsibilities, the *Suvidha-Kendras* were fully computerized, and networked to a central server dedicated exclusively for the Mission. *Suvidha* Kendras were oriented to look after the needs of the whole family, including children, adolescents, youth, senior citizens, differently-abled, etc., besides women. The original GRC concept which were largely centre based and functioned as "Stand alone units" earlier started providing services to those girls/women who could come out and seek livelihood training, non-formal education, basic health services available in these centres. The *Suvidha Kendras* in respective districts were aligned to the district centre and each district centre became the district hub for information and gradually taking up the role of a clearing house too, linking the *Suvidha Kendras* to different departments of the government.

Emphasis was on the most vulnerable people like the homeless, women and children headed families, families involved in vulnerable occupations like rag pickers, casual-daily wagers, etc. The district centres mentored, and monitored the work of field level NGOs which involved direct intervention in the areas of Health, Nutrition, Non-Formal Education, Micro-finance, and facilitator's role like creating awareness about government entitlements and facilitating linkages to programmes/services of different governments, including identifying the school drop-outs and mainstreaming them into regular schools.

The MC programme took a measured step in the direction of inclusive e-governance with the launch of the System Integration (SI) Application for streamlining social services delivery of selected schemes. A central repository of database of all vulnerable families was created at the PMU level. All the District Centres and *Suvidha*

Kendras were e-enabled and connected to the central portal. This enabled the centres to access a common database of vulnerable population surveyed (though conventionally referred as the 'beneficiary', in the MC a new term of "Entitlement holder" started being used for such persons who were eligible for government schemes/programmes) for providing access to various 'benefits' and provisions.

The SSS society/MC became recognized as a registrar under the unique identity project and Mission Director, MC in Delhi had a headstart as a registrar *vis a vis* other states gaven the preparedness in terms of database, enrollment centres located in the community in the form of *Suvidha Kendras,* with provisions of required IT hardware and software and trained professionals already in place in each of the centre. What further ensured readiness to take on the challenge of UID was the previous experience of RSBY enrollment which these centres had been exposed to (after lot of initial hesitation from the Labor department whether this could be done by centres managed by NGOs), and successfully demonstrated their strength in handling the process of community mobilization, enrollment and the required follow up. Women from the community were especially trained to handle the process of enrollment, getting the RSBY card made and then assisting other women to avail the health package in private hospitals which the health insurance card provided.

Benefits of Information Technology

The information technology infrastructure/foundations in the Mission was used to develop a very rich and useful data-centre, and a mechanism to map what a person would be eligible for given the vulnerability parameters and the scheme eligibility across nine departments, namely Health and Family Welfare, Education, Women and Child, Social Welfare, Food and Civil supplies, Labour, Urban, IT, and SC and CT/OBC/Minority/backward classes. Parameters of vulnerability involved socioeconomic conditions like senior citizens, differently-abled people, economically weaker population, widowhood, single and deserted women, women headed household, those suffering from HIV/AIDS or chronic illness, etc. The use of IT was intended to counter the problems in the existing system of, duplications, delays, multiple windows for the citizens, multiple criteria, and repetitive process of filling same information in multiple forms for different departments and even within the same department for different schemes. The system could not detect this earlier because the records were manually maintained in the Department of Food and Civil Supplies that allowed frauds to be detected.

MC came up with simple business process engineering to merge over 40 different application forms for applying to various welfare schemes in just one form which could capture the unique information about the person once and the same could be computerized and used for transmission to different departments, as and when required.

The much needed simplification of government procedures, increase in inter-departmental coordination and information sharing, and enhancing the community information about the system was made possible by intelligent use of IT in the programme. A reliable and dynamic database of households, with unique identification system, was developed so that it could be used by all the line departments for sanction and delivery of social services. A strong IT backbone was envisioned under the Mission at its very nascent stage in order to provide technological solutions for checking duplications, and improving the efficiency, transparency, and accountability of service delivery. The need was to put in place a reliable mechanism to identify those who were entitled but not drawing their benefits, and reach out to them in a systematic manner.

The *Suvidha Kendras* played a pioneering role in creating a common, dynamic data base of beneficiaries (who were called entitlement holders) which could be accessed by all line departments delivering social benefits. The effort involved the local community in identifying and authenticating the applicants. In the process, the *hitherto* unreached and the most vulnerable sections of the society could be brought into the system. Introduction of IT facilitated automated processing of applications, tracking applications, thus infusing transparency, and accountability in the system. Earlier there was an artificial crisis created in availability of forms, since the same had to be obtained from the office of the MLA or the central office of the department of social welfare. By putting these forms on the website, the system now had an easy access, and reduced the unnecessary trips which had to be made only to access the forms.

Foundation for Effective use of IT

IT for masses and not classes drove the MC to invest in human resources, hardware and software on a scale which was unprecedented. Demystification of technology enabled women to handle things easily through simplified, transparent and user friendly processes which were customized to suit people and not the government department. Empowering resource poor women to use and avail IT helped overcome the socio-cultural barriers formerly precluding women's equal access to opportunity in the areas of health, education, skill, social protection provisions, etc. *Suvidha Kendras* allowed access to women and girls who are otherwise reluctant in approaching the government departments so that the information and knowledge portal at each centre would serve as means of public information for them. Women could go to the centre and find out what programmes they could get enrolled into, for which a customized mapping software was developed and provided to each centre.

The *Suvidha Kendras* becomes the first interface for the people of the community to respond to their multiple needs starting with earmarking of a help desk cum information desk which is easily visible with adequate display of IEC material, mapping of resources in the area, with digitized database of the vulnerable population living therein, and the sight of young girls sitting behind the information desk with host of information in the computer became a visible sign of empowered women. Outreach strategies to reach the most vulnerable families, using tools like Mobile kiosks, organization of camps for getting the community enrolled for different certificates like disability certificate in which the women community mobilizers and SHG outreach worker, information desk staff played an active role and made it possible to bring services to thousands at the doorsteps and women in particular benefitted from the programme. Women mobilizers with laptops in the field or behind their PCs were trained on information of all government schemes, their procedures, conditions and formalities, guided the local community members to complete the forms and other documentary requirements for availing services. At the same time they were given the mandate to verify the authenticity of information through home visits, and update the digital record as may be needed. Discrepancies had to be reported to the next supervisory officer. The digital record of all the vulnerable households in the catchment area earmarked for the GRC-SK became a newfound tool of local area planning and right targeting. Applications for different services could now be made online using the computerized facilities at *Suvidha Kendras*. A robust Grievance Redressal Systems was also set up at each centre. Each form required a proper documentation of all records which were to be forwarded to the district level for further processing after scanning the records. Digital tracking of applications was now made possible in the new system. The chain of command involved submission of records to the district resource centre which in turn would take up these cases with the District convergence forum headed by area DC and the respective line department. Electronic submission of MIS reports was made mandatory in the prescribed format.

IT Use at *Suvidha Kendras*

Realizing the large volume of data and information associated with the welfare schemes of nine departments of the State Government, a robust Information Technology system was vital to facilitate coordination between different actors. E-Governance initiative was built in the programme right from the conceptualisation of MC. *IT enabled Suvidha Kendras, helped to build a strong intelligence system for maintaining a record of the unserved/underserved even though most vulnerable, and meeting all the conditions for availing government welfare programmes/other service, and also ensured that such persons could be included in the planning process and get covered under different programmes/services. They also helped to map the services given by different departments/ bodies of the government with reference to each vulnerable family who had been captured in the Mission's database for better targeting of the schemes.* The range of services mapped included mainstreaming of children in 6-14 age group, since after identifying the drop out or out of school children through household survey and community mapping, the GRC-SK would make efforts to follow up on their admission with the government schools in the vicinity. MC database and the System Integration project thus immensely helped the government in meeting its legal obligations and social commitments.

The implementation of System Integration Project was with a view to carry out the necessary GPR (Government Process Restructuring) of the institutional processes and mechanisms deployed in the sanction and disbursement of various welfare schemes, and develop a robust web-based application to be used by all line departments. The integration had to be preceded with creation of a dynamic database of potential beneficiaries / or entitlement holders. The participating departments could draw-from this common database of the beneficiaries, avoiding duplication of efforts, and increasing the reach of delivery of social services, especially to those most marginalized. The System Integration project facilitated automating the processing of applications and providing application tracking, thereby increasing transparency, efficiency and objectivity of departments' internal operations.

The master database was created based on extensive field surveys carried out in association with community based organizations in three phases, during 2008-2011. A large number of women were trained to do the mapping and data entry, while developing full ownership in the local community for the database thus developed. All scheme facilitation and implementation points were IT enabled and connected to a central server to facilitate the flow of information. This also provided at the scheme delivery points, instant access to the beneficiaries' socioeconomic data, thus making verification easier and eliminating hassles related to submission of various forms and documentation for availing these welfare schemes.

For the implementation of System Integration Project, M/s Vayam Technologies Ltd., was awarded work in October 2009. They were responsible for developing software/ applications for about 45 social welfare schemes implemented by 9 line Departments (including Social Welfare, Women and Child Development, Food and Civil Supplies, Urban Development, Health and Family, and Revenue Department, etc.). As part of the implementation of System Integration Project all *Suvidha Kendras,* run on a day to day basis by credible civil society organizations, selected through a transparent process involving field appraisals and interviews to judge their capacity, got computerized and got connected to central database through web-based applications.

This network of *GRC-Suvidha Kendras* was able to create a single registry system of 40-odd welfare schemes of government through a single convergence application providing an interface and access to eligible people for a particular scheme. The initial vulnerability index based survey was digitised and centrally maintained at the Project management Unit (PMU) after a strong process of field authentication. The database was actively used by the PMU for planning future interventions. This unique identity provided by MC actually became a precursor to the UID and demonstrated the advantages of such applications at a time when this had not been started anywhere else in the country at this scale.

The business process engineering and IT foundations, including the demonstrated readiness of the GRC-SKs laid the foundation for initiating the "Aadhaar Enrolment" on the MC platform. The beneficiaries' unique biometric based

BOX 1

Services provided through the new business process reengineering of the community nodes- *GRC-Suvidha Kendras*

- ✩ *Identification of the unreached that invariably are the most deserving, vulnerable sections of the society.*
- ✩ *Involving the community based organizations working at grass-root level in authentication of applicants.*
- ✩ *Automate processing of applications and provide application tracking, thereby increasing transparency, efficiency, and objectivity of service delivery.*
- ✩ *Creating a common, dynamic data base of beneficiaries accessible to all line departments, thereby avoiding duplication of efforts, checking leakages, and increasing the reach and speed of delivery of social services.*

identity, needed for the '*Aadhaar*' programme was dovetailed with the MC database. In Delhi, the Unique Identification Project (Aadhaar) was launched on 2nd October 2010 with the Hon'ble Chief Minister handing out the first *"Beghar Card"* to the Homeless, with the help of which they were able to enrol for Aadhaar. For the remaining vulnerable population residing in the vulnerable pockets of Delhi, MC took a conscious decision of once again taking the services to the doorstep of the community. These IT enabled, *Gender Resource Centre-Suvidha Kendras* (GRCSs) were found equipped and fit to take up the role as enrolment centres for UID, especially since any enrolment first required community mobilization, which was a forte of the CSOs, as demonstrated successfully during the household and homeless survey, RSBY enrolment drive and even earlier during the Stree Shakti camps.

BOX 2

Unique Identification- AADHAR- some key features

- *To make enrolment for Aadhaar simpler, Mission authorized GRC staff as "introducer" so that those living in the catchment area but without any standard Proof of Identity or Proof of Address could still be easily brought within the ambit of the universal identity system.*
- *The centres covered almost all clusters of vulnerable pockets for a large reach of the population.*
- *Over 2.3 lakh vulnerable were enrolled and approximately 48000 Aadhaars generated till 2011. The data correction exercise was also integrated with the UID enrolment process. This opportunity was utilised to allow corrections in the Mission's initial survey generated database.*
- *This exercise brought the kendras on a common platform and enabled the centres to access a common database of the vulnerable population surveyed for providing various services, and access to schemes.*

Another scheme called *"Dilli AnnashreeYojana" (DAY)* is an example of the successful use of central web-enabled database system for empowerment of women. DAY has been recently launched by the Govt of NCT of Delhi for providing food subsidy of Rs. 600 in cash to the Most Vulnerable/Vulnerable Households (MV/V HH) not covered under BPL cards and *Antodaya Anna Yojana* (AAY). The scheme is being implemented by MC with the support of its *Suvidha Kendras*, even though reflected as one of the schemes of Department of Food and Civil Supplies since the budget comes from this department. The money is transferred in the bank account of the senior-most female member of the households. This scheme is totally IT based scheme which includes online data updation, verification and approval procedure. The foundation of this scheme was laid by the survey of the vulnerable households which happened in 2008-09, and its quick implementation was made possible only because of the strong readiness of MC programme, in which IT systems already set in place played a key role in the cash transfer mechanisms.

The GRC-SKs which are centres run by NGOs through funding support provided by the government (approximately 15 lacks per annum) were responsible to locate the potentially eligible households using the data base of vulnerable household already created under the MC. Pre-filled application forms on the basis of available data were supplied to these centres by the MC head office, including the address and contact details of the potential beneficiary household, their demographic profile and bearing the social and occupational vulnerability that one or more members of the household may have had. The GRC-SKs were to visit the household and record if it faced any vulnerability during that time and also to verify if the household had become a recipient of subsidized ration for poor (i.e the BPL ration) in the public distribution system (PDS). If the household met the requirements of enrolment in the DAY, then the GRC-SKs were to advise the household to get their application form recommended by the area Member of Legislative Assembly (MLA). The functionaries of the GRC-SK are also expected to facilitate the required authentication processes at all levels. Once recommended by the area MLA, the form is to be sent back to the GRC-SK for onward submission to the Food and Supplies Officer of the area for the latter's decision on sanction of the cash entitlement. The entire enrolment process is closely monitored by the PMU of MC and the food and civil supplies office, including timely interventions with banks to redress the problems in getting bank accounts opened in the name of women member of the household in whose name the cash had to be transferred.

BOX 3

Dilli Annashree Yojana (DAY)

- ☆ *The DAY is a cash transfer scheme which provides a fixed amount every month directly to the bank account of the female head of the beneficiary household. The cash amount corresponds with the subsidy in the ration provided to Below Poverty Line households. The current amount is six hundred rupees per month per beneficiary household. The beneficiaries include the spatially, occupationally and socially vulnerable households who do not benefit from the ration provided to the BPL/AAY through the Public Distribution System.*
- ☆ *One of the unique features of DAY is that the bank accounts opened for all the beneficiaries are linked to the Aadhaar number of the Unique Identity Authority of India which makes it possible to eliminate duplicity of enrolment of beneficiaries in the scheme. The scheme benefits from the huge backend technological activities of UIDAI for Aadhaar enrolment which is a unique number given to every citizen based on their biometric records.*

Electronic reporting and monitoring system was developed in which the GRC-SKs are required to enter online the details of the enrolment related steps such as – whether the form was filled up, whether the required documents were collected from the beneficiary, whether the form was approved by the GRC-SK and handed over to the household for getting MLA recommendation, whether the form was recommended by the MLA and sent back to household and then back to the GRC-SK, whether the form was forwarded by the GRC-SK to the area FSO and whether the claim was finally sanctioned by the FSO. The GRC-SKs are to enter all these information on a daily basis which is then monitored by the Nodal Officers and other the senior management at the State Level.

Unlike few other schemes which put onus on the part of the beneficiaries to open up bank accounts, the DAY facilitates opening up of a no-frill bank account/s (called basic savings account) by inviting leading banks whose system is UID linked. This requires cross mapping of Gender Resource Centres and location of bank branches and then allotment of one or more bank branches to each GRC-SK for conducting weekly camps for the eligible beneficiaries to submit the application forms for opening of bank accounts at the GRC-SKs. Technological solutions to the difficulties of managing the large number of beneficiaries with small number of Community Mobilisers (CMs) and bank officials were developed. This involves the grouping of eligible beneficiaries based on their area of residence, possession of Aadhaar card number and the recommendation of the area MLA.

The beneficiaries are invited to the GRC-SK through local level interface by field functionaries, or through telephone/SMS etc. (since majority of the beneficiary households have at least one mobile phone). This helps in saving time of the beneficiaries and better management of the process of filling up of the forms, by facilitating the completion of relevant documents at the bank account opening camps. The list of the final beneficiaries as approved by the FSO with the details of the bank account number, IFSC Code of the bank for electronic transfer, *Aadhaar* card number etc. are shared with the Lead bank which acts as the Sponsor Bank for the scheme for transfer of the benefits through the National Payment Corporation of India. Thus, an integrated technological platform serves as the backbone for implementation of this unique social protection scheme *i.e.* DAY with the final outcome being a regular source of cash in the accounts of woman head of the household- certainly a step towards her empowerment.

Other Interventions for Women Empowerment through *Suvidha Kendras*

Women are central to these *Suvidha Kendras* established under MC Programme with focused interventions designed for their economic, social and psychological empowerment. The Woman Empowerment Component is aimed at holistic development of the marginalized vulnerable women through interventions in the areas including Literacy, Health, and Livelihoods. Under a well-structured programme, each of these centres provide Non-Formal Education, Vocational Training and Skill Development, Health and Nutrition through camps and clinics and also to instil the practice of thrift and micro credit through formation of Self-Help Groups (SHGs). Promoting community participation to bring about positive attitudinal and behavioural changes towards issues of women and girls has been fundamental to MC Programme.

Training and Capacity Building

Training and capacity building of women for livelihood generation followed by giving them entrepreneurial activity and has become one of the important initiatives. Women are trained in cutting and tailoring, embroidery, beauty culture and computer education free of cost at these centres. Women SHGs/other women groups have been given opportunity to operate the Canteen at the Delhi Secretariat as a model to stimulate more such enterprises. The canteen caters to about 2000 persons on an average. Though, the primary objective was to provide wholesome, hygienic food at reasonable costs, the focus of the initiative is also on empowering women in the process by means of training and employment opportunities. The canteen being successfully operated by Stree Shakti NGO not only provides front end services but also backed by techno-savvy women. This has already inspired many other women groups to constitute Self help Groups (SHGs) and start profitable micro-enterprise activities. Recently at the historical Teen Murti library, another unit of Stree Shakti SHG has started running a canteen in the series of many other similar offers pouring from various organizations.

Convergence of Resources

The huge investment in Information Technology, including the hardware, software, MIS development and training was possible due to linkage with the national E-Governance programme. The IT department of Delhi Government provided financial support to the MC initiative to strengthen the IT base. The World Bank also undertook a programme for providing the blueprint of strong IT architecture. With the UNDP CCT programme, the data analysis could be given a more robust picture. The USAID supported training of community functionaries involving hundreds of women in *Rajiv Gandhi Shahri Beema Yojna (RSBY)* enrolment process.

Challenges

Any successful IT initiative and E Governance programme entails multi-pronged efforts to lay strong grounds and the right foundations which can create an effective "systems" architecture. MC hence has been a journey in which simultaneously there have been attempts to rationalize policies, simplify procedures, and create sensitive response mechanisms closer to the community. The human touch has been taken in as much an earnest manner as the technological application. The challenge has been to keep the desired pace and align the pace of progress on the various fronts. The concept was novel and initially there was lack of confidence amongst many that all this was too ambitious and could never succeed beyond a partial demonstration of intent for a short while. It was also thought that once the main drivers namely the then Chief Secretary and the key lieutenants were changed this programme could never sustain. The initial scepticism however started giving way when down the line several champions emerged and the ownership grew. Initially the participating departments were not ready for computerized environment. Lack of IT skills amongst the functionaries engaged by the NGOs from the community was also a major challenge faced while implementing the MC project. Besides the basic infrastructural challenges, there was a resistance to change from status quo and perception that the initiative was impinging on the 'sovereignty' of the participating departments.

The transparency ushered by the use of IT was not always received with appreciation from all stakeholders involved. Aligning the pace of multiple activities was a huge challenge since delay in one dimension would affect the take-off of another activity. The smart card initiative planned for each person surveyed and potentially eligible for government services had to be revisited in view of parallel developments of

RSBY card, and UID project. The system of de-duplication/authentication process planned in the original architecture could not materialize at the pace initially envisaged. Other implementation challenges included the currency of two systems in the transition phase.

Future Prospects

The evolutionary nature of the programme and its flexibility, instead of being cast in stone to follow a straightjacket approach, remains both its strength as also its weakness. MC has many elements to it which are like building blocks. The blocks can be stacked in different ways to give different shapes to the building. The real challenge would always be as to who inhabits these spaces created by these blocks. If the support to those intended, *i.e.* the real poor, vulnerable and technology which is an important pillar of this building continues to empower the poor, the disadvantaged women, then this model can influence many other governmental initiatives the world over striving to meet similar goal/s. For its unique efforts MC received the prestigious Commonwealth Association for Public Administration and Management Award 2010 for Innovation in Government Services and Programmes winning the overall Gold category, and the UN Public Service award in the following year. However, the programme has yet to see its fullest potential being realized. For this it is important that the processes and early learnings are revisited as we progress, and there is always a system of policy feedback loop to respond to the need of mid-course corrections as and when merited in any aspect of this gigantic programme.

Acknowledgement

The support of *Dr. Sunita Sangar, Senior Research Officer, National Resource Centre for Women for putting this paper together is acknowledged with deep appreciation.*

References

1. Cecchini Simne and Christopher Scott, 2003, Can information and communication technology application contribute to poverty reduction? Lessons from Rural India Information Technology for Development, 10: 73-84.
2. VAPS 2004. Enhancing women empowerment through Information and Communication Technology- A Report, by Voluntary Association for People Service (VAPS), Tamil Nadu, Submitted to the Department of Women and Child Development, Ministry of HRD, Government of India.
3. Kenny, C. Navas- Sabater, J. and Quang, Z., 2000 ICT and Poverty, World Bank, Washington DC.
4. Kenny, C. 2002 Information and communication Technologies for Direct Poverty Alleviation: Costs and Benefits, Development Policy Review, 20 (2), 141-157.
5. Torero M. and Von Braun J., 2005 ICT: Information and communication Technologies for the poor, International Food Policy Research Institute (IFPRI), Issue Brief No. 40.
6. World Bank, 1998 World Bank Report, New York, Oxford University Press.

Some of Useful Links for more Details

1. Case Study: Mission convergence: http://negp.gov.in/pdfs/Case per cent 20Study_Mission per cent 20Convergence.pdf *accessed on 07.02.13*
2. The Gender Resource Centre - *SuvidhaKendras* of Mission Convergence: A case studyhttp://www.itforchange.net/sites/default/files/ITfC/MC_fieldnotes.pdf, *accessed on 07.02.13*
3. SamajikSuvidhaSangam- A status report: http://www.missionconvergence.org/A per cent 20status per cent 20report.pdf, *accessed on 07.02.13*
4. Mission Convergence 2013 http://www.missionconvergence.org/women-empowerment-component.html, *accessed on 07.02.13*

Chapter 15

Empowering Women in Developing Countries through Information Communication Technologies

Geeta Malhotra[1] *and Yashpal Malik*[2]

[1]*Country Director, READ India*
97, 3rd Floor, Sector 23, Dwarka – 110 077, Delhi, India
E-mail: geetamalhotra@readglobalindia.org
[2]*Programme Officer, READ India*
134, 1st Floor, Sector-2, Rohtak – 124 001, Haryana, India
E-mail: yashpalmalik@readglobalindia.org

ABSTRACT

This paper explores the role of ICTs in empowering Indian rural women, through a review of ICT initiatives in India by READ centers. Among the varied tools, the knowledge centers and the community radio were found to have the greatest potential in reaching women with locally relevant content.

There is immense potential for ICTs to create new employment opportunities for rural women and to contribute significant gains in efficiency and effectiveness in rural women enterprises. While ICTs can play an important role in empowering rural women, women's access and use of ICTs and empowerment clearly depends on the vision and operational agenda community and grass root workers. Therefore, strengthening the ICT initiatives of such key players can go a long way in empowering rural women. Besides generating locally relevant content and enhancing the capacities of rural women in accessing ICTs, efforts are also needed to bridge the different types of digital divide (rural-urban; men-women).

The growth in digital technologies during the last decade has been astounding. If we look at the information scenario with ICTs, one starts to think of the ideal situation where in:

- People in all walks of life are empowered to seek, evaluate, use, and create information effectively to achieve their personal, social, occupational and educational goals.
- All people have access to information services in whatever form they present themselves.
- Information resources relevant to local communities are accessible, available and also affordable.
- Digital access (computers, internet) Mobile phones being used for information creation and access.
- People with computer literacy (ICT skills) and media literacy.
- Facilities for storing and preserving information whether in analogue or digital form in all communities.
- Online search tools available to all, including multilingual searching; Open access to information and resources.
- New literacies (information, computer, media) incorporated into education curricula.

The information should be available in local and simple language, so that everyone in the community, including women, can understand and make the best use of it. There is a need for initiation of a knowledge revolution through effective and meaningful use of ICT. Participatory communication is essential to achieve a common vision of the society for development.

Keywords; *Women empowerment, Information communication technology (ICT), Rural development, Gender development, Communication, Skill development, Poverty alleviation, Digital divide.*

Introduction

In an age of media and information boom, even though rapid and fast exchange of information and knowledge sharing is possible, yet the fact that many communities still remain distant and removed from such global knowledge networks, is an important and poignant reality of the times with ICTs becoming boom in the 21st century. There is a need to take ICTs to the rural communities and enable them to use the tools for effective communication and as a means for livelihoods.

Empowerment of women in the context of knowledge societies entails building up the abilities and skills of women to gain insight into the issues affecting them and also building up their capacities to voice their concerns. It entails developing the capacities of women to overcome social and institutional barriers and strengthening their participation in the economic and political processes so as to produce an overall improvement in their quality of life.

Access is the central issue necessary for women's empowerment. Women have traditionally been excluded from the external information sphere, both deliberately and because of factors working to their disadvantage such as lack of freedom of movement or low levels of education. ICT opens up a direct window for women to the outside world. Information flows to them without any distortion. This leads to broadening of perspectives, greater understanding of their current situation and causes of poverty and the initiation of interactive processes for information exchange.

Access to ICTs is crucial if they are to be means for women's economic empowerment. We need to work towards universal access. It is important not only to establish physical facilities, such as communication networks or computers, but to ensure that these facilities are utilized by their users to the greatest possible extent. Women's access to and use of ICT is constrained not only by technological infrastructure, but also by socially constructed gender roles and relations. According to a UNESCO report on Gender Issues in the Information Society, the capability of women to effectively use information obtained through ICT is clearly dependent on many social factors, including literacy and education, geographic location, mobility and social class.

In developing countries like India, more than 90 per cent of women work in the informal sector and also in rural areas. These women engage in economic activities such as handicrafts and sewing, weaving of baskets and fabrics, working as vendors, and working without any contracts or benefits. These are the women who need and deserve poverty alleviation programmes more than any other. IT will expose these women to telecommunication services, media and broadcast services that will create markets for their products and services. The challenge will be to reach these women and provide them with ICT tools that they feel can make a difference in their income generation potential.

Some examples from the initiatives taken by READ (Rural Education and Development) India while setting up Community Library and Resource Centres in rural India and encouraging women to take part in the income generating programmes at these Community Centres. Further details can be known from the hyperlink "http://www.readglobal.org" "www.readglobal.org".

READ Centres provide a safe space to women to share their perceptions, learn traditional and about new innovations using Information Communication Technology tools, basic computers, radio, video, digital stories etc. and become active users of the technology provided at their door steps followed by well-planned training programmes.

Background of Women in READ Centres

One such Centre has been set up in West Imphal in the village called Taboungkhok, where the Maitie women are born entrepreneurs. Given the opportunities they spend most of their time focusing on better earnings for the family. The following are some highlights during the programme interventions in 2008-2010:

- ☆ Maitie women work with commitment and concern. They weave for the family needs and also for getting some financial support from the products. They are receptive to all the initiatives of alternative development. However most of them are not willing to pursue any of the income generation activities as full time activities because they have to look after their family, agricultural farms, worry about cultural barriers and responsibilities and much more.

- ☆ Meitie women, thus, are able to earn reasonable income from the part-time activities for their petty requirements, such as pocket money for their children, mainly for the girl child, for providing hospitality to the visitors at home and the relatives, for the ritual obligations, for garments, cosmetics and essential items. Such part-time income gives them respect and decision making capacity in the family and society.
- ☆ Most of Meitie women take their surplus products such as vegetables, grains, art and craft objects, etc., to the Imma Market or Mother Markets designed mainly for the women in various localities in the State, and sell them. Recently, Government of Manipur has built two special huge markets for women to provide them with better space for better earnings. Mrs Sonia Gandhi inaugurated these markets.
- ☆ Several women are working in the schemes of NREGA and earning some money to support their families financially.

Access to technology can open a new arena for their market products. Presently, the products are only seen by the local women, who buy, but if, with ICTs, they are able to reach other markets, it will be a boon for them.

Another case study is from Saharia Community in Bhanwargarh village of Kishanganj block, Baran district, Rajasthan.

Jan Jagriti Gyan Kendra (READ Centre) is working with women who are having Gooseberry processing as one of the sustainable programme for the Centre. Women are producing gooseberry products and selling them not only in Baran but the products go to Kota market too.

The tribal women have approached the schools for providing gooseberry candies during the mid-day meals to address anemia in the school children. Studies have shown the improvement in hemoglobin level in the malnourished children in selected participating Anganwadis and primary schools of the block.

These women are making their financial records on tally and a few of them are using computers, but still to develop market linkages. The process of development takes time, but once women are exposed to ICTs and acceptability and accessibility is there, they carve their own path forward.

Another case study from the Geejgarh Gyan Kendra (READ Centre) in Geejgarh village, district Dausa in Rajasthan revealed that the local adolescent girls and women are keen to learn computers, stitching and library skills to manage the Community Library and Resource Centre and would like to promote the Apparel Centre in their village for better marketing skills and linking them with the export houses, and other commercial hub in Rajasthan.

The project interventions enabled the women to understand the importance of Information Technology tools and how these tools can help them not only for their own empowerment but the empowerment of the community as a whole.

The key challenges with regard to ICT infrastructure identified included:

- ☆ Power (lack of reliable supply)
- ☆ Maintenance of equipment and infrastructure, as well as timely repair of errors and faults in the equipment
- ☆ Availability of operating system software's
- ☆ Integration of new technology with the existing and well-proven classroom techniques
- ☆ Appropriate training/orientation
- ☆ Lack of bandwidth and other technical support functions across states
- ☆ Institutionalizing the process of capacity building with open schools/ universities

Materials and Methods

After five years of interventions with the community, a study has been conducted in all five READ centers in the States of Manipur, Rajasthan, and Delhi. The five centers in these States have been chosen because READ is doing ICT interventions in these areas since 2010. The sample size was 125, *i.e.* 25 women respondents from each center.

To evaluate the impact of the ICT programmes on women, a detailed evaluation was done. The evaluation included rigorous analysis of how well the project methods and activities met the diverse needs of the participants and the aim of empowering women using ICTs. Multiple research methods were used for assessing the impact:

- ☆ Participant observations of project activities such as workshops and online conversation groups.
- ☆ In depth, semi-structured individual interviews with twenty-five participants from each project location were done with the adolescent girls and women.
- ☆ Focus Group Discussions were conducted with the group at each center to find outthe impact of ICT tools on their lives.

After the data was collected from the field, it was processed using a computer through the use of Statistical Package for Social Science (SPSS) and Excel. Data was collected from various parts of rural India, in Manipur, Rajasthan and semi-urban village of Delhi inthe age group of14 to 30 years.

Results and Discussion

All READ centres have an ICT section having computers, internet connection and a qualified trainer to train women on use of ICTs. As the centres are within the vicinity of the village, women come to the centre easily after doing their house-hold jobs, and quite often come two-three times in a day. They find the centres as convenient place to share and learn. The following are available at the Centre:

1. More detailed teaching and more time for hands on experience or practicals
2. Availability of Hindi Typing facility
3. Results of exams, PAN and passport status checking
4. Forms available for government schemes like birth, death, differently challenged people and old age pension, etc.
5. Teacher is from the village itself and understands the local challenges
6. Internet to download the needed information and for children games and educational material, etc.
7. Customized course curriculum
8. Course completion certificate from READ and Government

Respondent Profile

From five READ Center's, a total of 125 women respondents were interviewed. In each center, the total number of respondents were 25. All the respondents were women those who had taken training on ICT tools from the centers and included adolescent girls and students, etc.

Educational Profile

ICT and Women's Life

It is true that ICT made life easier for both men and women. When asked whether the introduction of ICT made life easier, 57 respondents viewed that they strongly agree, followed by 57 who somewhat agree. Only 5 respondents said that they do not know or cannot say whereas only 1 respondent said shedoes not agree with the idea.

Access to ICT

The rapid expansion of IT has reached all sections and almost all areas of India. Though, rural and remote areas still lagged behind as compared to urban areas. Figure 15.1 shows that 56 respondents said ICT is available easily in their area. But about same number, 50, said that they somewhat agree that it is available easily.

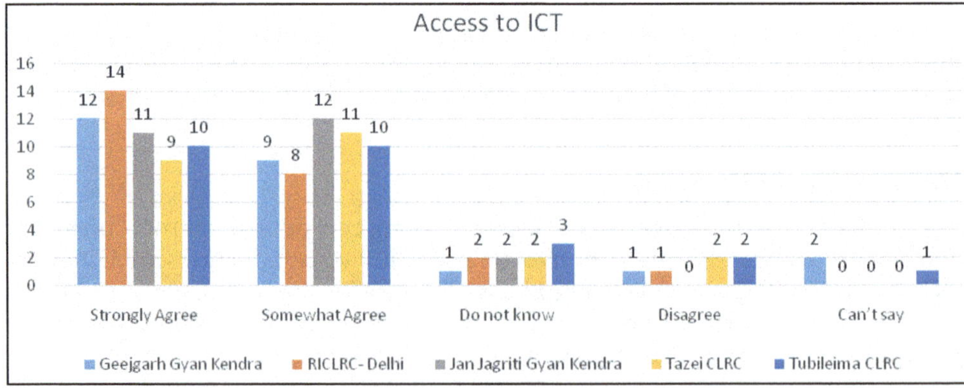

Figure 15.1: Access to ICT.

6 respondents disagree with availability of ICT in their area. 10 said they don't know and 3 had no opinion. The reason for this is low literacy rate, the value of ICT is not realized, and, a few not interested as there is no need for ICT in their lives.

Use of ICTs

The ICTs are not limited to a particular purpose. It enabled its users for the variety of purposes. Out of the six options presented to the respondents, for the purpose of the use of ICT, 105 respondents said they use it for the Information and Learning. It made clear that most of the ICT users use for the purpose of information sharing. Only 8 respondents said they use for the communication, 3 for the booking of tickets, 7 to get information in different aspects of life and 1 person said they use it for the purpose of banking and insurance. Out of 25 respondents, if we get the response on all five usages, it means that the women are not ignorant. If constant exposure and training is provided, they can very well accept and use ICTs.

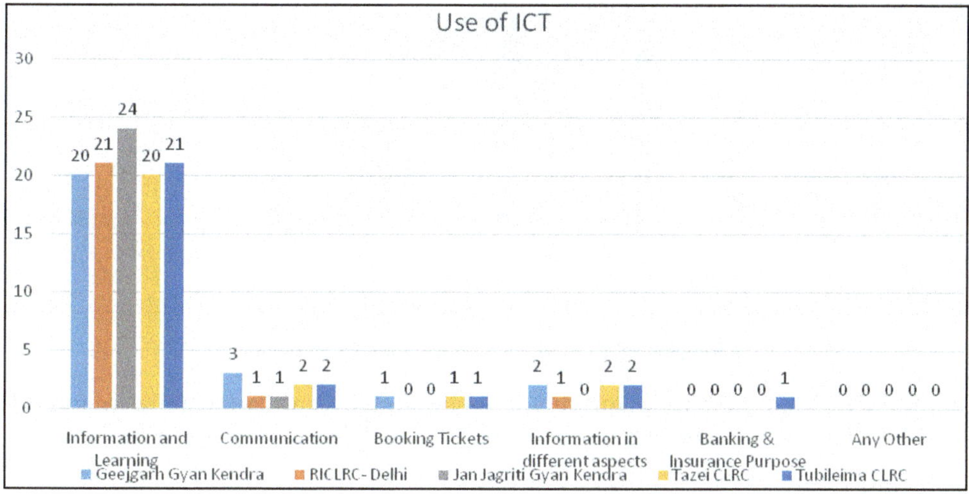

Figure 15.2: Use of ICTs.

Access to Internet

In all parts of the world, internet is easily available and a good number of populations use it very easily. Women also have easy access to internet facility. Out of the 125 respondents, 97 said that the internet is easily available; whereas 28 respondents said access to the use of internet is not easy.

Sources of Access

In India, there are also a variety of sources of the availability for internet. The respondents who said that there is easy availability of the use of the internet, the highest 64 said they prefer READ Center. According to them, it is easily available in their locality. 34 said they use it in the Cyber Cafe. 22 persons who use internet at their home, stated it was cheaper to use at home instead of other places. Only 5

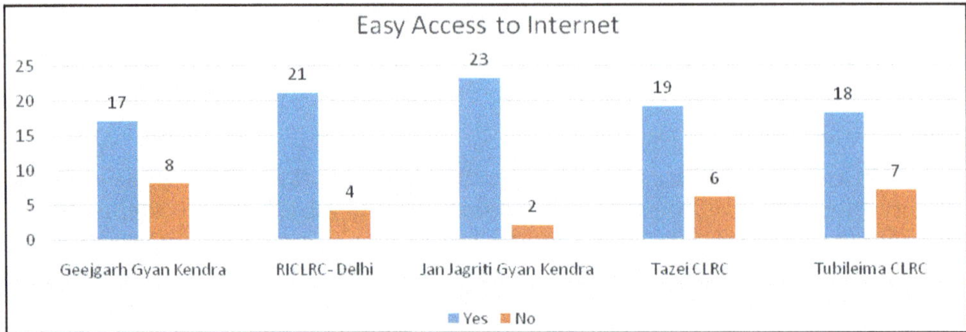

Figure 15.3: Access to Internet.

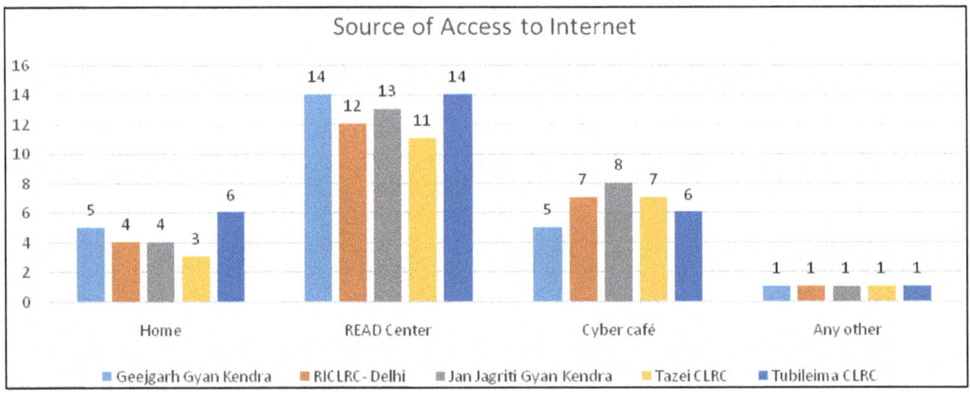

Figure 15.4: Sources of Access to Internet.

respondents said their use of internet is not limited to a particular source of internet instead they use it according to the means of availability and need of the hour.

Internet User

Internet is also easily available to its user. When asked whether the introduction of internet helped them to get information easily, nearly 58 respondents strongly

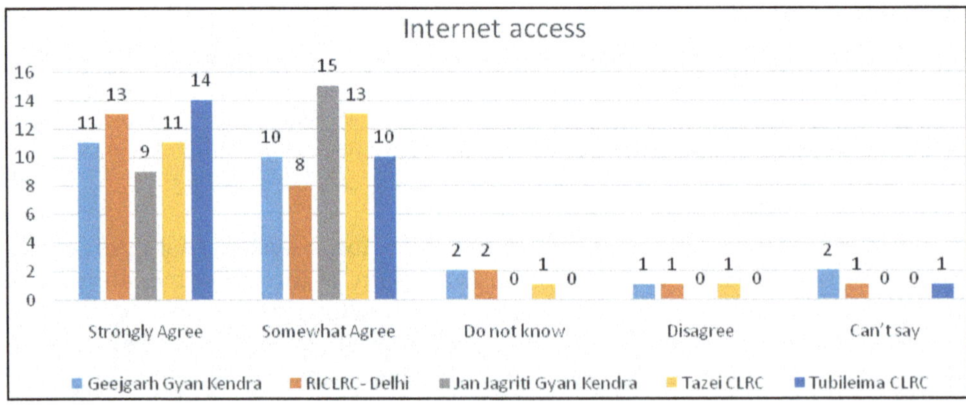

Figure 15.5: Internet Access.

agreed and also about the same number 56 agreed somewhat with that idea. It can be seen from the Figure 15.5 that the use of internet is more familiar/accepted in certain areas.

Purposes of ICT Use

The respondents use internet more frequently for learning purpose in comparison to other uses of the internet. They use it to gain information which is more-easier and quicker than any other source like newspapers and books. 88 respondents said they use internet for the learning purposes whereas only 17 said they use it for email purposes. They also use it for study purposes like getting various materials for their study related subjects. Though, it is a small number, 11 to use internet for study purposes but this is more familiar among the students.

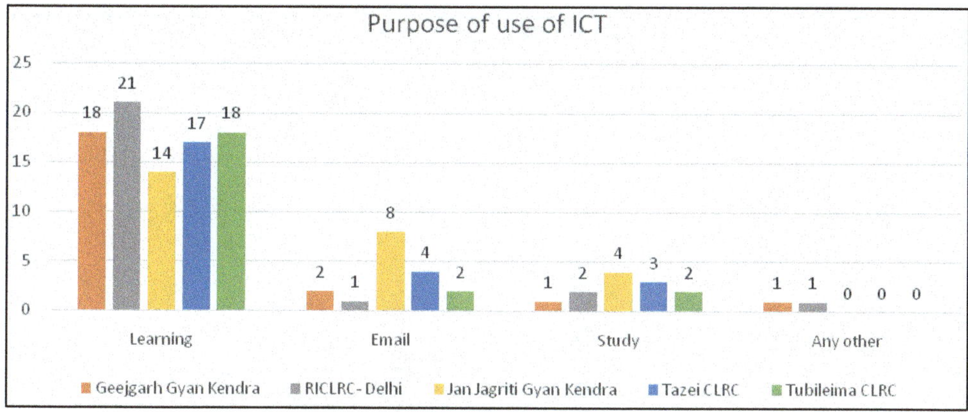

Figure 15.6: Purposes of Use of ICT.

Freedom of Decision Making in Family Matters

The access to ICTs have enabled them to build self-confidence, the interaction with others who have not been exposed to ICTs gave them respect and dignity as their knowledge is recognized, especially in rural areas, when one says that they got the particular information on the internet or the data base available at the READ Centre.

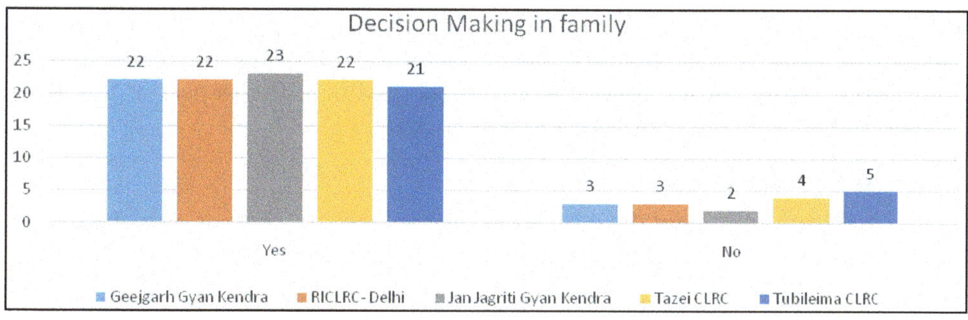

Figure 15.7: Decision Making in Family.

Their regular visits to the READ Centers gave them the recognition in the village and the family members have also recognized their potential, resulting in involving them in decision-making processes for their families.

The study found that a large number of women, 110 participate in the decision making of their household activities. Only 15 said that they are still in a position where they are subservient to their male counterparts in the family.

ICTs have Helped them to Learn and Know what is Happening in the Country and Out Side

With the introduction of ICTs at the centers, 47 respondents said that they are able to know what is happening across the globe more easily, though 33 respondents agreed only somewhat with this statement.

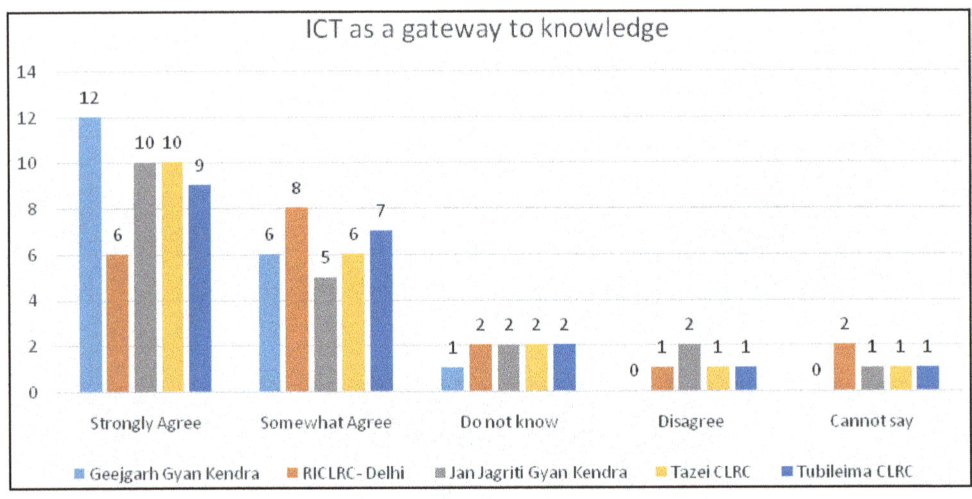

Figure 15.8: ICT as a Gateway to Knowledge.

Expectations from READ Centers

Women are now expecting some higher courses like hardware and networking, other than the basic course they learn from the Center. They say it is good to have a basic course to learn for beginners especially for villagers it would be a boon, as many villagers would have a first time opportunity to learn computers. For the benefit of the students who have completed the basic courses and for some students who opted computers as their major studies in their college due to the impact of the READ Center, and thus higher courses were more in demand.

Recommendations

- ✰ Keeping in view the education system in India, there is a strong need to provide access to computers at the school level creating awareness on importance of ICT.
- ✰ School Management Committees should be made conscious of this important role.

Figure 15.9: ICT Activities.

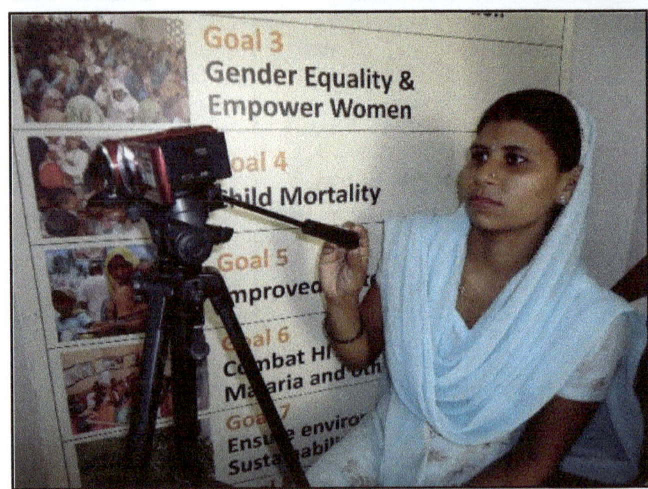

Figure 15.10: ICT Activities.

- ☆ School teachers should play an important role in providing access to computers to children.
- ☆ Panchayats should be made aware of not only providing infrastructure but also on effective use of the infrastructure.
- ☆ Education is a state subject. Under 29 responsibilities given to Panchayats, ICTs could be a part of the responsibility along with strong capacity building of teachers.
- ☆ Use of human resources available at the rural level volunteering their time on training and capacity building.
- ☆ ICTs do make a difference, as it enables access to information, offers opportunities at an individual and community level. ICTs makes it easier for women to accomplish things faster but it does not always lead to empowerment.
- ☆ While welfare and access issues of empowerment may be addressed by ICTs and to a certain extent increase participation; there are limitations to how much ICTs can actually do.

Conclusions

An increase in self-confidence regardless of their age or occupational status has emerged as a significant unintended outcome of the ICT enabled READ Centers. 96 per cent of rural women either "strongly" or "very strongly" felt that their self-confidence had improved due to the initiative. 88 per cent of women told that they are able to take decision in their family because of knowledge gained from ICT programmes. 91 per cent of the women respondents either "strongly agree" or "somewhat agree" that ICT has made their lives easier.

References

1. Enhancing Women Empowerment through Information and Communication Technology by VAPS, India http://www.wcd.nic.in/research/ict-reporttn.pdfý
2. Empowering Women Through ICT, Spider ICT4D Series No. 4, 2012.

   http://www.comminit.com/ict-4-development/content/empowering-women-through-ict.
3. Gender equality and empowerment of women through ICT.

   www.un.org/womenwatch/daw/public/w2000-09.05-ict-e.pdfý
4. ICTs and Empowerment of Indian Rural Women.

   http://www.crispindia.org/docs/4 per cent 20CRISP per cent 20Working per cent 20PaperICTs per cent 20and per cent 20Empowerment per cent 20of per cent 20Women.pdf
5. ICT for rural prosperity.

   www.mobileplus-india.net/wp-content/./Pawan-Kumar-MP-Pres.pdfý

6. ICT and Women Empowerment.

   panegov.net/oegf/./06-ICT per cent 20 and per cent 20Women per cent 20Empowerment.pdf

7. White Paper. Information and Communication. Technologies (ICT) in Education for Development.

   unpan1.un.org/intradoc/groups/public/documents/./unpan034975.pdf

8. Rasheed Sulaiman V, N J Kalaivani, Nimisha Mittal, 2011, ICTs and Empowerment of Indian Rural Women, India.

9. Women and ICTs; Suman Jain.

Chapter 16

Women's Empowerment through ICT: Case Study of 2 Projects under Ministry of Science, Technology and Innovation (MOSTI), Malaysia

Puspadevi Kuppusamy

ICT Policy Division, Ministry of Science, Technology and Innovation,
Level 5, Block C5, Complex C,
Federal Government Administrative Centre,
62662 Putrajaya, Malaysia
E-mail: k.puspadevi@mosti.gov.my

ABSTRACT

The role of Information and Communication Technology (ICT) has increasingly been recognised as a tool of women's empowerment. The usage of ICT can provide comprehensive instruments for women for improving social relations, providing job opportunities, seeking information and getting ideas on solving issues pertinent to women. ICT becomes a momentum for well-being of women in various aspects towards improved health, social and economic development. In this presentation, two projects will be discussed that empower women organized by agencies under Ministry of Science, Technology and Innovation (MOSTI) Malaysia. The first one is called "1nita" which is an initiative under the Malaysian government's 2nd Economic Stimulus Package. The project was implemented to provide a sustainable platform for women entrepreneurs and women-owned SMEs to build business capacity for the future. This project, run by.my Domain Registry, an agency under MOSTI aims to empower women in achieving long-term business success by taking advantage of technological developments and the Internet. Project 1nita is an e-business platform strictly for women-led SMEs and entrepreneurs to enhance their capabilities and skills in ICT, while providing them with a platform to conduct business online. The second project is the Internet Entrepreneurship Programme (IEP). It is a programme organized by the Multimedia Development Corporation (MDeC), an agency under MOSTI launched in 2009. The objectives are to spur the development of entrepreneurs from rural and semi-urban communities nationwide, to increase their marketing capabilities and skills using ICT and Social Media. Target group of this programme is the micro and small enterprises led by women. This entails employing new and innovative approach in helping existing businesses to grow their market share. These two projects for empowering women through ICT can lead to knowledge production, innovation, social transformation and thus, equitable sustainable development.

Keywords: *Women, ICT, Empowerment, 1nita, IEP, Domain registry, MDeC*

Introduction

Information and Communication Technology (ICT) comprise a complex and heterogeneous set of goods, applications and services used to produce, process, distribute and transform information (*Ramilo; Hafkin; Jorge, 2005*). The term ICT has been used to encompass technological innovation and convergence in information and communication, leading to the development of so-called information or knowledge societies, with resulting changes in social interaction, economic and business practices, political engagement, education, health, leisure and entertainment (Ramilo and Villanueva, 2001).

ICT plays a vital role in today's world. ICT is a necessary tool for the development of a nation especially developing countries. Over the past decade, there has been a growing understanding that these technologies can be powerful instruments for advancing economic and social development through the creation of new types of economic activity, employment opportunities, improvements in health-care delivery and other related services, and the enhance networking, participation and advocacy within society (Ramilo and Villanueva, 2001).

ICT could give a major impact to the economic and social empowerment of women. The usage of ICT provides comprehensive instruments for women for improving social relations, providing job opportunities, participation, increasing freedom to seek information and getting ideas on solving issues pertinent to women. This can then become a momentum for the well-being of women in various aspects towards improved health, social and economic development.

Materials and Methods

ICT has much to offer for empowerment of women in terms of access to information and services. They could save lives (nutrition, health), opportunity to secure new jobs in the knowledge economy (media, web, programming, data entry, sales), women-friendly working models (can work from home, telecommuting), especially in the cultures which do not allow women to circulate freely and mix with men in the office. Besides, these, ICT offer access to education at all levels and at all times through e-learning - beyond what is offered in the face to face traditional schooling system; access to micro-credit, possibility to make financial transfers (remittances); and provides a voice (through email, creation of web sites, chat rooms, distribution lists) for women in ways that never existed before.

This paper will discuss two projects that empower women in terms of economic upliftment by agencies under the Ministry of Science, Technology and Innovation (MOSTI) of Malaysia. Malaysia is a member country of the NAM S&T Centre and MOSTI is the focal point of the NAM S&T Centre in the applicant's country. Hence, the two projects are discussed that empower women organized by agencies under MOSTI.

1nita

The 1nita Programme is an initiative by MYNIC Berhad, an agency under MOSTI. 1nita Programme provides an e-commerce platform strictly for women-led SMEs and entrepreneurs to enhance their capabilities and skills in ICT, while providing them with a platform to conduct business online. This e-commerce enablement will help women-owned business organisations make their products and services accessible to a larger market and expand their customer reach. The 1nita initiative is an innovating approach for women entrepreneurs to do business by going online. 1nita aims to empower women in achieving long-term business success by taking advantage of technological developments and the Internet.

With innovation being a key component in the development of a high-income, knowledge-based economy, Programme 1nita is in line with the Government's efforts to introduce a new economic model for Malaysia. Prime Minister Datuk Seri Mohd. Najib Tun Razak's call to Malaysians is to utilise the Internet and web applications to generate new and additional sources of income.

A dedicated e-commerce platform, www.1nita.my, has been developed by MYNIC Berhad. It is a platform for women involved in SMEs to conduct their business online and expand their marketing horizon beyond the geographical boundaries through Internet and availability of technology-based platform. The key objectives for the Programme 1nita include:

- ☆ To empower women-owned SMEs in taking advantage of the technological developments and join the global economy and market spawned by the Internet, towards increasing business potential and revenue.
- ☆ To help women-owned business organisations in making their products and services accessible to a larger market and expand their. customers reach.
- ☆ To enable women-owned SMEs in achieving the objectives of the new economic model.
- ☆ To provide skills and knowledge of building a successful online business through trainings and workshops. The aim is to help participants achieve competency in developing website, upload and update contents and transact online; enrich the information of their company, products and services; reap measurable benefits from the web within a year.

- ☆ To build a cohesive support community of women entrepreneurs that will go on to expand the 1nita initiative through networking, training and mutual assistance.
- ☆ To establish a robust e-market place exchange through which women could promote lucrative e-commerce opportunities amongst Malaysia women entrepreneurs.

1nita Programme incorporates a comprehensive approach on e-commerce development, business development, entrepreneurial skill, as well as personality building for the participants involved. The module for the programme is a combination of the following:

- ☆ E-Commerce – 3 days training course on the 1nita platform (www.1nita.my) which includes registration of domain names, website development and knowledge on application and operations of the platform.
- ☆ Entrepreneurial Skill – Short courses in effective Internet marketing and retail/wholesale business opportunities.
- ☆ Personality Building – Professional image grooming.

The 1nita Programme has gathered participants from across Malaysia in numerous workshops that were held from its launch in 2010. As a result, at end of 2012, 1nita has successfully positioned 1661 women from SME's in the Internet domain by providing them with.my domain names and online presence via the 1nita platform.

For the year 2013, MYNIC Berhad continues to carry out the mandate to implement the 1nita initiative under the purview of MOSTI. The targeted number of workshop participants for 2013 is **1800** from several states in Malaysia namely Sabah, Perlis, Johor, Pahang, Kelantan, Perak, Kedah and Pulau Pinang.

Improvisations were made to the 1nita Programme such as addition of new courses to the modules and improvement of the existing 1nita platform. There is a plan to embark on an aggressive marketing drive to further promote the 1nita Programme throughout various conventional and social media platforms. The presence of participants should be ensured continuously who have been involved in the programme since 2010, through various retention approaches as well as continuous improvements made to the 1nita platform. New features will be incorporated from time to time to the training module and the e-commerce platform to ensure the sustainability and continuity of 1nita in the future.

Internet Entrepreneurship Programme (IEP)

The Internet Entrepreneurship Programme (IEP) is a programme organized by the Multimedia Development Corporation (MDeC), an agency under MOSTI which was launched in 2009. This programme was launched to spur the development of entrepreneurs from rural and semi-urban communities nationwide, in order to increase their marketing capabilities and skills using ICT and Social Media. The target group of this programme is micro and small enterprises led by women. It is a quick win strategy for development and enhancement of capacities and capabilities of micro and small enterprises. This entails employing new and innovative approaches in helping existing businesses to grow their market share; in addition to fine tuning conventional approaches such as merger and acquisition, new policies, government procurement and direct foreign investments (DFI) assistance.

One of the approaches of IEP is to encourage micro and small enterprises to adopt technology, in particular the Internet, in order to achieve 'Internet Presence' through various online tools and platforms *e.g.* Social Media. The IEP enables women to benefit from Internet Marketing activities either via expanded market reach, or increased visibility and enhancement of company profile.

Objectives of the IEP are to facilitate business owners to create online presence and start marketing their products and services online, to impart necessary skills and tools to participants in familiarising them with online marketing, and to increase usage of Telecenters in community as the access and economic focal point for that community, wherever applicable.

MDeC has collaborated with various NGOs, trade and business associations, regional development authorities, government agencies and even ministries in delivering the IEP. Amongst the collaborating partners involved are: Association of Bumiputra Women Entrepreneur Network of Malaysia (WENA), Union Malay Economic Action Council (MTEM), Dewanita, Yayasan SALAM, Association of Bumiputera Women Entrepreneurs Malaysia (USAHANITA), Association of Bumiputara women in business and profession Malaysia (PENIAGAWATI), Association of North Region Women Entrepreneurs (USAHAWATI), Malaysian Association of Bumiputera ICT Industry and Entrepreneurs (NEF), Malay Businessman and Industrialist Association of Malaysia (PERDASAMA), Economic Corridors of Malaysia such as Iskandar Regional Development Authority (IRDA), Northern Corridor Implementation Authority (NCIA), East Coast Economic Region Development Council (ECERDC), Agencies such as People's Trust Council (MARA), Small and Medium Enterprise Corporation Malaysia (SME Corp), Development Authority of Terengganu Tengah (KETENGAH), Sarawak Economic Development Corporation (SEDC), Ministry of International Trade and Industry (MITI), etc.

IEP activities are as follow:

- ☆ Briefing and Profiling session
- ☆ One-on-one Interview session
- ☆ Workshop on Internet Marketing (2 days)
- ☆ Workshop on Social Media (2 days)
- ☆ Online Coaching and Mentoring (6 months)
- ☆ Tracking and Monitoring (6 months)
- ☆ Impact Assessment (after 6 months)

Programme Content is as listed below:

- ☆ Identification and listing of Keywords via External Keyword Tool
- ☆ Principles of Search Engine Optimization (SEO)

- ☆ Strategizing your Internet Presence
- ☆ Product Photography Technique
- ☆ Listing on Google Places
- ☆ Free portal listings (Classifieds, Business directory, Mini-website)
- ☆ Vertical portals/specialized sites: forums, FB groups/pages (listing and advertisement)
- ☆ Photo marketing and sharing
- ☆ Knowledge/Information Marketing (via document, presentation, video, photo)
- ☆ QR Code as marketing tool
- ☆ Software as Service (SaaS) business management tool
- ☆ Managing online presence
    - i. Understanding the characteristics of each platform and how to manage with regards to ease of updating, content strategies, listing validity, etc
    - ii. Scheduling for updating and content improvement
    - iii. Further marketing on other suitable platforms
    - iv. Monitoring effectiveness of each platform/tool
- ☆ Planning and Managing Impact to Business
    - i. Preparing existing operation to be 'online ready'
    - ii. Anticipate and plan for expected increase in sales in terms of production, operation, financing, HR, etc.

Results and Discussion

1nita

Figure 16.1 shows the number of participations for 1nita Programme of 2010-2012. At the end of 2012, 1nita has been successfully delivered to 1661 new online women entrepreneurs in accordance to the nation aspirations. Targeted participants of 1800 are shown for the year 2013. It shows that more women entrepreneurs are being approached and trained. By the end of 2013, it is targeted to have 3461 domain names ending with.my for the women entrepreneurs. The 1nita programme has helped to increase the market for small and medium enterprises of women besides equipping women with ICT skills.

Internet Entrepreneurship Programme (IEP)

As of now 25 IEP workshops have been conducted in 13 states in Malaysia ever since the launch in 2009. Meanwhile 500 women entrepreneurs were trained and ICT enabled. Table 16.1 shows the direct contribution of 500 website/social media for their products. According to the feedback received, 30 per cent of the participants said that their revenue has increased by at least 8 per cent and 500 new jobs were also created.

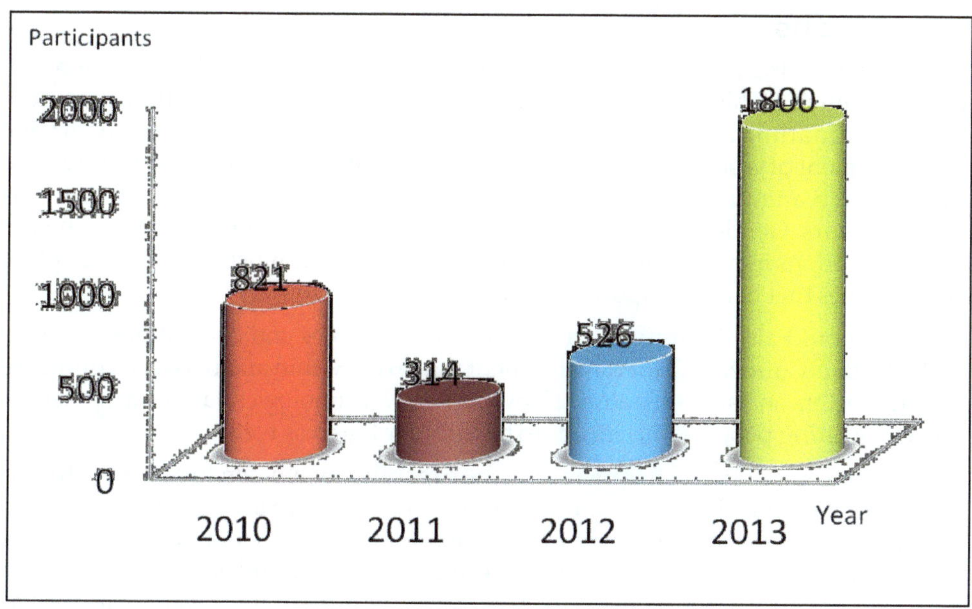

Figure 16.1: Chart Showing Participation of 1nita Programme According to Years.

Table 16.1: Increment of Sales of Participants

Sl.No.	*Company Name*	*Industry*	*Per cent Increment of Sales (After Programme)*	*New Job Creation*
1.	Puteri Sharon Resoucers (Sambarunding)	Food Industry	40 per cent	1 permanent staff Additional staff is hired when needed.
2.	Sanggul Depeterana	Wedding Planner	30 per cent	7 permanent staff and 3 temporary Staff
3.	Atikah Songket	Clothes/Textile	20 per cent	4 permanent staff and 1 temporary staff
4.	Kay Bakery	Food Industry (Bakery)	45 per cent	2 helpers (part time)
5.	Karaz Synergy SdnBhd	Food Industry (Drinks)	50 per cent	1 Helper
6.	Waha Hills Trading	Service	60 per cent	6 staff (part time)
7.	Ehsan Murni Foods Trading and Consultancy	Food Industry (Frozen)	45 per cent	1 business partner
8.	Laina Design	Printing	20 per cent	1 assistant designer
9.	Din Beramboi Resources	Clothes/Textile	40 per cent	1 sales assistant
10.	NB Enterprise	Food Industry (Bakery)	60 per cent	2 permanent staff
11.	S.F Zas Trading	I Food Industry (Frozen)	70 per cent	4 permanent staff and additional staff is hired when needed.

Conclusions

ICT have a big potential for promoting and achieving sustainable development, however much of this potential is yet to be realized. Adequate allocation, resources and coordination among ministries responsible for ICT is essential for the empowerment of women. An enabling environment at the national level requires that policies give attention to promote ICT among women in rural areas as well as entrepreneurs. Gender perspectives have to be taken into account in identifying the ICT implications in policies in all sector areas. Governments must commit in adopting programmes that help to empower women.

We should also identify the differences in resources and capabilities to access and effectively utilize ICT for development that exist within and between countries, regions, sectors and socio-economic. However, new technologies have a vast potential for empowering people which needs to be fully exploited as well.

Advancement in ICT has increased the capability of women to reach out to and to connect with the global community. However, it needs to master and leverage on the transformation taking place with the technology, to ensure that we maximise the positive impact of ICT that brings to our economy, society and environment. These measures are crucial for women to boost the economic competitiveness as well as societal well-being.

References

1. www.1nita.my (accessed 26 May 2013).
2. www.innovasia.com.my/ý (accessed on 26 May 2013).
3. *Ramilo C., Hafkin N., Jorge S., 2005.* Women2000 and Beyond, http://www.un.org/womenwatch/daw/public/w2000.html (Accessed 26 May 2013).
4. Adoption, 2003.United Nations Commission on the Status of Women Ramilo.G., Villanueva. P., 2001. Concepcion.

APPENDIX A : 1nita

Photos during 1nita Workshop in Kota Kinabalu, Malaysia (April 2013)

Briefing Session

Discussion among participants

Participants are doing the hands on session

Participant receiving Certificate of Attendance

APPENDIX B: Internet Entrepreneurship Programme (IEP)

Photos during Internet Entrepreneurship Programme (IEP) across states in Malaysia

Briefing Session

One to One Coaching Session

During Hands on Session

Participant Receiving Certificate of Attendance

APPENDIX C: Success Stories of IEP

IEP Pahang : NB Enterprise

www.nb-kekcoklat.com.my

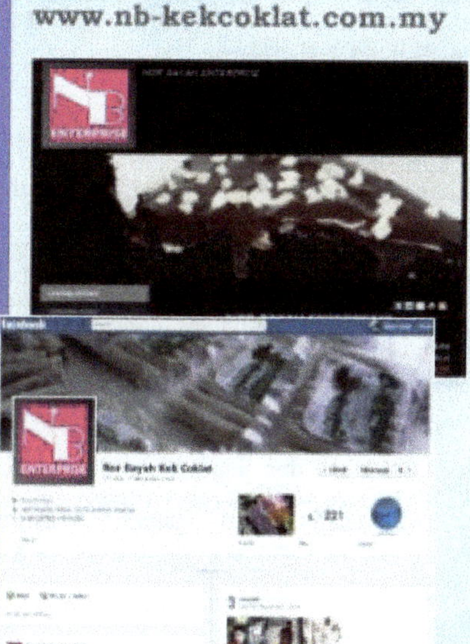

- Produk utama –Kek Coklat telah dapat dijual diluar Kuantan dan kini sudah sampai ke Johor, Perak, Kuala Lumpur, Negeri Sembilan dan Terengganu.
- Kebanyakkan tempahan diperoleh melalui Facebook, dan bisnes lebih menonjol melalui Facebook. Purata peningkatan jualan kini mencecah sehingga 90%.
- Boleh dikatakan mendapat tempahan dari pejabat-pejabat sekitar Pahang hampir setiap minggu. Juga mendapat pertanyaan dari pelanggan di Indonesia dan juga New Zealand.
- Telah memperluaskan perniagaan dengan mengadakan latihan , kursus dan peluang kepada para pelajar bagi latihan industri yang menjumpai kami melalui laman web dan Facebook.

IEP KL : Laina Design

www.lainadesign.com.my

- Tempahan meningkat kerana pelanggan mudah untuk memilih design yang disediakan melalui website.
- Produk-produk juga turut mendapat tempahan hingga ke Singapura.
- Sedang menyertai Program GROW PKNS (Siswazah), diberi booth selama 6 bulan dan perlu cuba untuk achieve target didalam bines yang dijalankan.
- Dapat menjalankan bisnes 24/7, dan mampu untuk bergerak pada kadar yang lebih cepat dan bersaing dengan trend terkini industri.

IEP Johor : Waha Hills Canopy

www.wahahillscanopy.com.my

- Peningkatan jualan untuk sepanjang tahun 2011 adalah 60K, tetapi untuk tahun 2012, sehingga Jun sahaja, dapat mencecah ke 60K.
- Perkembangan yang membanggakan adalah melalui sumber online yang digunakan, Waha Hills Canopy berjaya mendapat tender daripada syarikat besar di Johor iaitu SWM Sdn Bhd.
- Seorang pelanggan turut memuji Strategi Pemasaran yang digunakan adalah amat baik kerana hanya menggunakan keywords 'Canopy Rental Johor', website Wahahills Canopy dapat dicapai, dan kini berada di halaman pertama carian Google.
- Jumlah kanopi bertambah dari hanya 4 buah sehingga ke 20 buah dan ingin mencapai target penambahan kanopi ke 30 buah. Pelbagai servis lain turut disediakan seperti *'underlayer with chandelier lamp'*, platform ,dan servis menyewa lori.

Chapter 17

Women in Research and Development Institutions: Case Study of National Center of Research

Eiman Elrasheed Eltayeb Diab, Migdam Elsheikh Abdel Gani, Maha Ali Abdel Latif and Gamma AbdELgader Osman*

Head of Department of Pollution and Environmental Studies,
Institute of Natural Resource and Environment,
National Center for Research, Khartoum, Sudan
**E-mail: eimandiab@hotmail.com*

ABSTRACT

The aim of the study is to evaluate the situation of women in the National Center of Research (NCR) in terms of women leadership positions, in relation to their academic qualifications as professor, associate professor, assistant professor, researcher, research assistant, technician and administrative staff. The structure of the National Center of Research consists of: general directorate, institutes, centers, commissions and units.

The study is based on the data obtained for the period of 1990-2012, which is divided into two sections with an approximate interval of ten years. The data showed that only 10 per cent of professorswere females in the first decade from 1990-2000, while 40 per cent professors in the second period of 2001-2011were females. It is noteworthy that none of the female staff were appointed as general directors ordeputy directorssince 1990. In the first ten year period only one woman held the position of a principal for a short duration of 2 years.

Women directors in different NCR institutes constituted 40 per cent of the total in the first decade and 85 per cent in the second decade. Among the head of departments, on the other hand, 50 per cent were women in the first decade and 90 per cent during the second period. Among technical staff in the first period women were approximately 75 per cent and constituted 85 per cent work force in the second period. As for theadministrative staff, currently the women constitute 90 per cent of the total staff force. Other department had undergone an even greater change in terms of women occupying various positions.

This paper discusses women in the NCR which need capacity building in leadership, skills, and awareness of the importance of role of women in higher positions.

The Role of Women in Research Institute

Introduction

Women play important roles in the society as procreators of the coming generation as well as producers of goods and services. In the modern economies, they also play important roles in the labour force (Onsongo, 2004). According to the European Sudanese Public Affairs Council Report (ESPAC, 2002), women in Sudan constituted slightly below half of the total population and form a critical portion of the human resource base. Regarding education, Sudanese women have enrolled formal educationsince the beginning of the last century. However, the first woman attending a university was admitted to the University of Khartoum in 1945 (Mohammed Saad, 1972). As an indication in supporting women education, the government of Sudan offers support to both sexes in education, so that the enrolment of women in formal education has been increasing yearly. hasgained a fairly high level in higher education, as well as administrative, academic and scientific research posts.

Scientific research in Sudan has been started since 1902 with the establishment of the agricultural research unit, in order to explore the possibilities of growing cotton under irrigation. The Welcome Tropical Laboratories were established in Khartoum in 1903, with the emphasis on medical research, but also conductedthe chemical and entomological research related to agriculture. In 1970, the National Council for Research was established to formulate and supervise the national research policy, and to fill in the gaps in research that have not been tackled by the existing research institutes. In 1992, the National Council for Research was converted into the National Center for Research (NCR) (Hassan, 2009).

The main research organizations are recently under the coordination of the Ministries of Science and Communication, Agriculture, Industry and Animal Resources. These organizations cover almost all aspects of science and technology. A few other specialized institutions are based in related ministries and private sectors (Hassan, 2009).

In order to promote women's participation in scientific research, and consequently in science and technology, it is very important to understand the pattern of employment and the reasons behind it. The resource of most of statistical data in this regard came from developed countries, and however, the shortage and unreliable data came from developing countries including Sudan. This situation makes difficult any quantitative estimation, and be a constraint in understanding the pattern and coming up with measures to correct it. Therefore, the objective of this study is to evaluate the status of women in research institutions, identify factors that may enhance, promote or inhibit the status and advancement of women.

Methodology

This study is conducted by data collection through research instruments including the interview guides and the documents analysis guides. The interview guides are used to interview key persons regarding the policies that govern employment, recruitment, appointment and promotion of staff to senior management position, and the possible reasons for the absence of women from these positions. The documents analysis guides include statistics and reports of research development and other official documents. The lists of researchers and technical staff are also used for gathering information about existing positions occupied by women in the research organizations. The obtained data are analyzed using quantitative and qualitative methods. The information obtained from the interviews and the documents are then analyzed through a process of content analysis and categorized into themes. Descriptive statistics, such as frequency counts of managers, and percentages of women managers are used to show the status of women in the research center.

Results

Table 17.1a summarizes the data concerning the numbers and the distribution of female and male researchers in different fields of science and technology in Sudan, while Table 17.1b shows the positions by women at different instututes of national Center for research (NCR) for two decades This data indicated that in all sectors of research, women proportion in all post degrees is less than men. However, it was clearly observed that the percentages of women increase with previous degrees. On the other hand, variation in women numbers among different fields is also observed. In some research disciplines, such as animal resources, biology and environment, the number of women is greater than man. However, in other fields such as atomic

Table 17.1a: Numbers and Distribution of Women and Men Researchers Across different Fields of Research in Sudan

Research Field	*Professors*		*Associate Professors*		*Assistant Professors*		*Researchers*	
	Female	*Male*	*Female*	*Male*	*Female*	*Male*	*Female*	*Male*
Agriculture	3	28	24	13	32	82	111	143
Animal resources	4	6	8	20	24	24	133	126
Biology and Environment	4	7	8	11	21	8	89	68
Atomic energy	0	3	0	4	1	11	38	64
Industry	0	0	9	10	15	5	12	16
Economics	0	0	0	1	0	6	8	9
Energy and geology	0	1	0	0	2	2	11	5
Information and communication	0	0	1	1	2	2	3	1
Total	11	35	50	59	97	140	405	432

energy, industry, energy, geology, information and communication, the proportion of women decreases as they rise up the science career ladder and is virtually zero at the top,although significant numbers of women gaining university degrees.

Table 17.1b: Positions Held (Percentage of the total) at the different Posts Reached by Women in the NCR Institutes 1990–2011

Positions Held of the Total Number of Researchers	*Duration*	
	1990-2000	*2001-2011*
Professors	10	40
Directors	40	85
Head of Departments	50	90
Technical Staff	75	85
General Staff	0	90

Distribution of the female and male researchers of the National Center for Research among different post degrees (professor, associate professor, assistant professor and researcher) was illustrated in Figure 17.1. It was indicated that the total number of women researchers attaining positions past degrees of assistant, associate professor and number of women researchers were higher than men, moreover, women professors represented 50 per cent of all professors in the National Center for Research. Figure 17.2 shows the position of women in the management system of the National Center for Research, as compared to men in position. The results revealed that despite the fact that the share of women researchers in key position of the research center (general director, deputy director and principal) almost zero, however, out of 12 Directors of research institutes were woman, and out of the 51 heads of department 27 were women. Figure 17.4 shows the position of women as technical staff in the National Center for Research. It was observed that among technical staff, out of 52 of Master

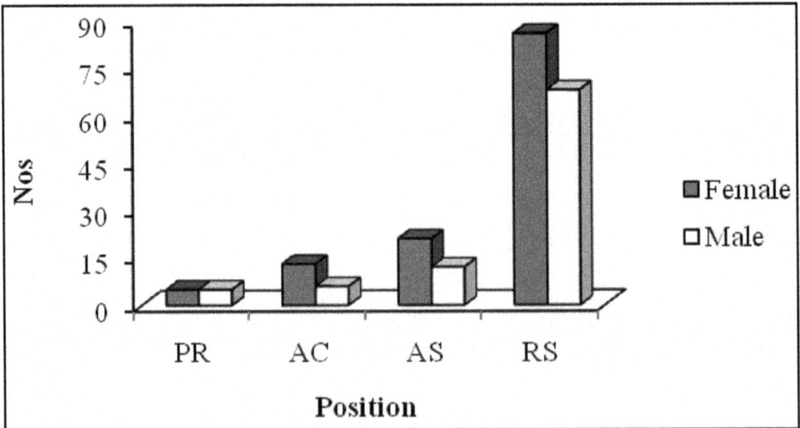

Figure 17.1: Distribution of Female and Male Researchers of the National Center for Research among different Post Degrees.

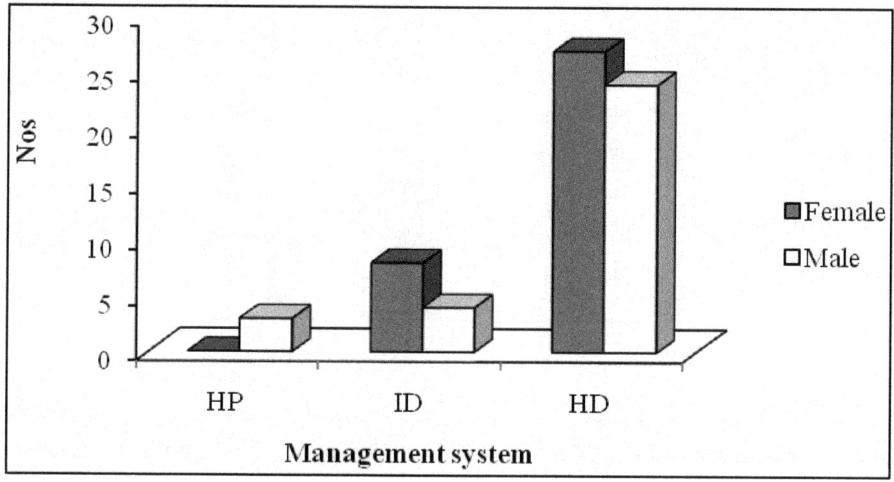

Figure 17.2: Situation of Women in National Center for Research Management System.

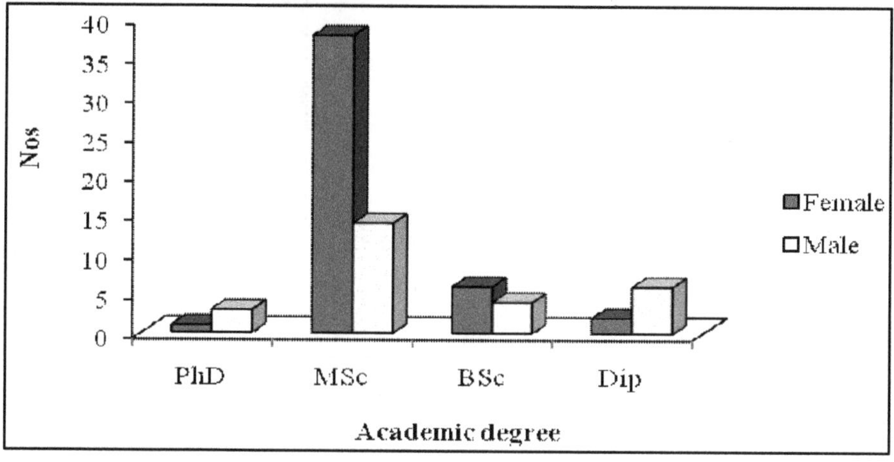

Figure 17.3: Distribution of Technical Staff in National Center for Research.

of Science holders, 38 were women. However, very few of women hold the Doctor of Philosophy degrees as compared to men.

Table 17.1a (replaced by 17.1b) determined the positions held by women from 1990 – 2011 at the institutes of NCR (NCR is composed of 9 institutes) as compared to men (in percentage)

Figure 17.3 illustrates a comparsion between women and men regarding qualification degrees (Diploma, B.Sc., M.Sc., Ph.D.) in NCR

Figure 17.5 Illustrates comparison between women and men regarding scientific prizes and awards in NCR

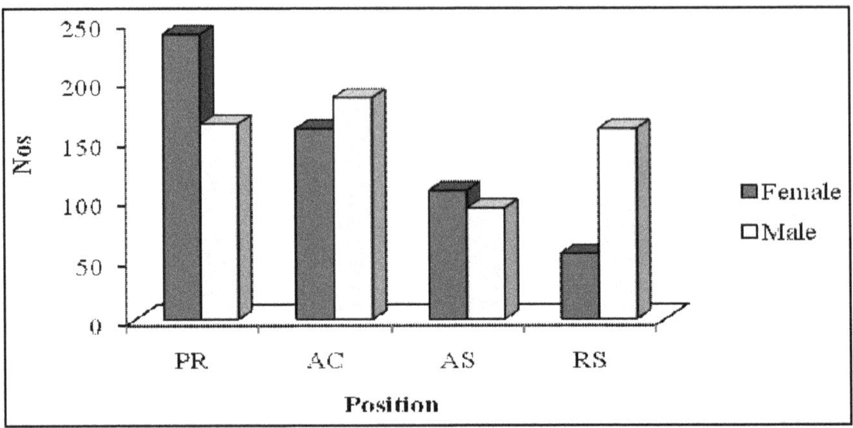

Figure 17.4: Contribution of Women in Publication of National Center for Research.

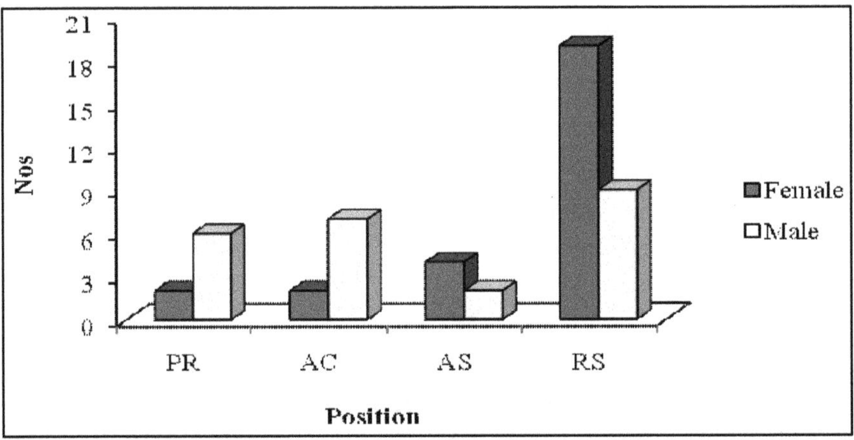

Figure 17.5: Prizes and Awards of Women in National Center for Research.

The scientific publications and scientific awards and prizes indicated that until 2011 the intellectual contribution of women in research stated that women researchers published more 580 scientific papers and represented more than 505 of total publications of the research center. Until 2011, women had also awarded 27 prizes and awards, which were more than prizes of men researchers.

To evaluate further the role of women in the institution management, both men and woman research managers were asked to explain their main responsibilities in the institution. The aim of this question is to find the extent of women are involved in the key decision-making responsibilities in research management. The answers to this question revealed that there are different responsibilities between women and men managers.

In the first interview conducted for women in the leading positions, the main focus of all was on the social obligation of women that consumes a significant part of their life. This is resulting in fewer female professors. There isactually no positive discrimination, since female professors would never take priority over the large number of male professors.In addition, it was revealed that 60 per cent of women reaching the age of retirement by the time they become Professors.

Discussions

As indicated by the above data, it is clear that women constitute a large proportion of research workforce in Sudan, particularly in the National Center for Research. The situation of women researcher in Kenya is better than many other developing countries. The study finds out the position of women occupying in the management level of university as to compare to those of men. The obtained data and information revealed that women occupya small percentage of the senior management positions (Onsongo, 2004). However, it is observed that the Sudanese women researchers have a particular challenge when they move up the hierarchy, as women have to compete with men.

Theproblem may be attributed to many factors. It was found that women are divided between two spheres: the management of the home and family, and the fulfillment of job responsibilities. The family commitments, either as the women's choice or as a result of cultural enforcement, have impaired women's capacity to meet their potential, and put them at a disadvantage in many science and technologyrelated jobs that are naturally dynamic and competitive (Taeb, 2005).

At the personal level, such factors as academic qualification (Ph.D.), administrative experience, management skills, confidence, assertiveness, high visibility, hard work and diligence, were found to enhance women's participation in research management. On the other hand, absence of these personal attributes were said to limit women's confidence in applying for position as senior managements.

Conclusions

Women produce huge quantities of publications, and consequently occupy their positions as heads of the departments and research institute directors.However, there is still a continuous decline when women move up the research ladder, despite the large numbers of women have higher university degrees. One of the issues that could be considered to solve the situation is to increase the age of retirement for females from 60 years to 65–70 years, because at the age of 60 people tend to be fit and can still deliver efficient activities.

References

1. ESPAC (2002).Women in Sudan. The European Sudanese Public Affairs Council.
2. Hassan, A. H. (2009). The Sudan National System of Innovation. The workshop on the Presidential Initiative on Reform of Science and Technology System in Sudan, Khartoum Friendship Hall.
3. Saad, M. (1972). "Notes on Higher Education for Women in the Sudan". Sudan Notes and Records 53:174-181.

4. Onsongo, J (2004). Factors affecting women participation in university management in Kenya Organisation for Social Science Research inEastern and Southern Africa.
5. Taeb, M (2005). Revisiting Women's Participation in Science and Technology Emerging Challenges and Agenda for Reform. UNU-IAS.
6. SAS (2010). Manual of reseachers under the umprella of Sudan Academy of Science fedral unioin Sudan Academy of Sciences, Sudan
7. NCR (2004). National Center for Research Manual, National Center for Research, Sudan (NCR) Reports.

Chapter 18

Globalizing India: The Sunshine Sector and its Shadows–The Call Center Industry: A Case Study

Rekha Pande

Professor and Head, Department of History, School of Social Sciences
University of Hyderabad, Andhra Pradesh, India
E-mail: panderekha@yahoo.com; rpss@uohyd.ernet.in

ABSTRACT

The present paper attempts to look at the so called sunshine sector, the Call Centres from a gender perspective, with a case study from Hyderabad, Andhra Pradesh. This paper is based on a case study with a sample data of 75 women who were interviewed over a period of one year in 2011-2012. We argue that the call centers present a very complex picture in terms of empowerment, agency and impact on the women working there. Information Technology is both an enabling and limiting factor. In ICT women are excluded by and large at the higher level and they are confined to the lower level in IT enabled services or call centres. The gender gap- especially the gap between men and women and how they benefit from Information technology has widened because women are less likely than men to receive technical education or be employed in technology intensive work. This is relatively true in rural areas among poorer population where women are already disadvantaged by their relative lack of education. While globalization has brought in many new opportunities and increase in income it has not really changed the gender division of labor in the workplace and at home or subverted patriarchal norms. Yet in some limited sense these women also have an agency for, as while there is a perception of very westernized liberated women, these women have a distinct Indian identity that is being continuously redefined and altered.

This paper focuses on the following aspects:

1. Globalization and structural adjustment in the developing world.

2. Engendering Globalization.
3. Historical background to the development of Information Technology in India.
4. ICT and constraints to women's participation.
5. India - The Call Centre Capital of the World and development of A.P. as hub of ICT.
6. The new realities of Call Center Employment- A Case Study.

Globalization and by extension information technology, play an important role in integrating markets and populations into a homogeneous web of consumers and producers. Opponents argue that globalization actually produces heterogeneous divides and exacerbates differences. Common rhetoric coins the twenty-first century as "the information age," embedded with a "knowledge economy," where globalization and IT is kin in the extended family of hegemony and pedagogy. It is the Information and communication technology or ICT, which becomes the flag bearer or the focal point of the globalization process. By collapsing time and distance ICT's enable individuals and organizations to develop networks of interests that transcend national boundaries. They support the pursuit of private interests that are independent of local and community needs. By transferring technology, resources and information from one region of the globe to another ICT's have also transferred practices and ideologies.

In the present paper we attempt to look at the information and communication technology in India with a case study of the Call Centres in Hyderabad in Andhra Pradesh and argue that the call centers present a very complex picture in terms of empowerment, agency and impact on the women working there. In ICT women are excluded by and large at the higher level and they are confined to the lower level in IT enabled services or call centres. The gender gap- especially the gap between men and women and how they benefit from Information technology has widened because women are less likely than men to receive technical education or be employed in technology intensive work. This is relatively true in rural areas among poorer population where women are already disadvantaged by their relative lack of education. Information Technology is both an enabling and limiting factor. While globalization has brought in many new opportunities and increase in income it has not really changed the gender division of labor in the workplace and at home or subverted the patriarchal norms. While there is a perception of very westernized liberated women, these women have a distinct Indian identity that is being continuously redefined and altered and one cannot escape the larger questions of capitalism that both enables and limits.

Globalization and Structural Adjustment in the Developing World

The 1990's saw a new phenomenon emerging in the world, referred to as globalization. This involved the spatial reorganization of production, movement of industries across the borders and the spread of financial markets that resulted in flexible productive methods and integration of production into global commodity and production chains. Businesses could treat the world as their field of operations and redeploy their capital and move the location of their production at will. There have been varied definitions of globalisation linking it to the arguments of the empire (Hardt and Negri, 2000), Clash of civilizations (Huntington, 1993) and Jihad vs. Mc World (Barber 1995). It is commonly understood as the economic integration and interdependence of countries and the move from traditional societies to a modern society (Featherstone, 1995). Structural adjustment becomes a very crucial aspect of globalization. Adjustment is often taken in response to severe macro imbalances which cannot be adequately controlled by normal policy instruments at the government's disposal. Adjustment would require sweeping changes in the organization and functioning of the economy. In order to tide over the financial implication of these changes, assistance from international financial institutions becomes an in separable part of this adjustment. Thus, when we talk of the reorganization of society in the aftermath of globalization, the restructuring of the production process was the prerequisite and the platform that helped this process.

It is widely recognized that in order for the world to be able to advance in the quest for sustainable development, peace, social justice, racial, ethnic and gender equality it is fundamental for every one throughout the world to have a greater participation in decisions that concern them and to develop their capacities. This possibility depends to a great extent on the access of these groups to the benefits offered by new and emerging communication technologies. Nonetheless access to these technologies is highly unequal in different geographical regions and social groups and this inequality contributes to increasing the gap between those who have access to abundant information resources and those who are deprived thus reinforcing the marginalization that already exists in terms of development and technical resources (Fatma Aloo, 1998). The whole idea that geography has become history is now turned on its head and was merely a hype.

While the information economy is flourishing and is now evolving a networking economy that is radically transforming the world of work, serious concerns have been raised if this revolution will be a powerful vehicle for gender equality or will an increasing number of women find themselves caught in the wrong side of the digital divide (ILO, 2001). Following the Okinawa Summit in 2000, the group of eight nations set up a digital task force to work on ways to eliminate the divide. UNO places the access to IT as the third most important issue facing women globally after poverty and violence against women (UN, 2000).

Engendering Globalization

In all the discussions on globalization and Information and communication technology, the gendering of this process is often overlooked. To many this is seen as a gender neutral technology. Feminist scholars have pointed out that Globalization has not changed the gender division of labor. The least skilled level of work with the lowest remuneration continues to be assigned to women following nearly universal gender division of labor and patterns of work organization. For poor women the existing inequalities and insecurities have intensified. Globalization has only widened gender disparities and increased the feminization of poverty (Pande, 2001). Some unskilled women have lost their livelihood as alternative sources produce goods cheaper and faster. For some women it meant a loss of rights, benefits and job security. Globalization has also led to the migration of a large number of women from developing countries. From the nineteen seventies an unprecedented number of women workers from developing countries entered both the formal and the informal labour force to service the global economy which has led to their increasing marginalization and pauperization (Pande, 2000).

Besides documenting the incorporation of women into the global economy, feminist scholars have also developed theoretical modes of explaining why and how this happened. Some have emphasized the masculinization of the development process (Boserup, 1970), structures of capitalism (Ward, 1984) to International gender division of labor with the intersection of global Capitalism with patriarchy (Nash and Kelly, 1983). These theories have pointed out that gender enters the globalization process in multiple ways. Gendering of the global economy involves not simply the division of women as workers but embodying the gendered principle through their ideologies, practices and policies. The impact of Information Technology on women's work in the context of globalization has been viewed differently by feminist scholars depending on where they are located. In developed countries, most of the literature on the impact of information technology on gender and work deals with the association of men with the technology and power (Wajcman, 1991). In developing countries women are looking at the issue not only in terms of gender relations with the men in their society, but also at western domination over innovation and as source of technology (Mitter, 1999).

Another area which feminists have studied is the global dynamics of work. Whereas earlier the focus was more on manufacturing jobs within the formal organizational settings, now there has been a broadening of focus of what is transnational, across organizations types to non-government organizations, information industries (Freemen, 2000). Gender employment patterns in the IT industry are already a cause for concern. The gender inequalities well established and documented in other sector of the labor force are already being replicated in the ICT sector. The number of women in IT jobs requiring advanced skills such as engineering and system analysis remains small. A large number of women are found at the lower level in IT enabled services. Even if women are able to acquire better education and training and begin to enter IT fields in great numbers, women's leverage in the IT job market may be undercut by feminization of labor. As computer based skills become more commonplace and the need for more workers to use them in a greater variety of ways grows, more women will be again recruited. But this will be at a lower wage because these will no longer be considered specialist skills, but merely something women can do (Reardon, 1999). The recent Law in America which stops Government from out sourcing its work in developing countries has further hit this sector negatively.

Acknowledging the multiplicity and diversity of women's narratives and experiences while exploring and deconstructing the categories of women and gender (Butler, 1990), academic feminists have vehemently challenged the scientist cloak and objective research. Feminists have turned towards agency and structure problematic with a feminist discussion of agency is increasingly emerging (Tina Basi, 2009).

Historical Background to the Development of Information Technology in India

Today a vibrant information technology industry has emerged in India. Government policy has also played a significant role in the creation and development of India's IT industry. Policies based on economic philosophy of import substitution during the seventies and eighties have given way to those aimed at liberalizing and globalizing the economy in the 1990's. Colonial rule had deprived India of an Industrial revolution. After independence India had adopted an economic policy that the public sector should expand rapidly and the private sector should play its role within the framework of the comprehensive development policy. In this strategy, private and foreign capital were strictly controlled by the government and not allowed to enter key industries reserved exclusively for the public sector. India's rejection of the neo- liberal approach to development and its opting for a mixed economy with State directed planning after independence in 1947 clearly reflected in the countries approach to high technology industries (Harindranath, 1999). During this period, multinationals IBM and ICL leased equipment often of obsolete technology to users in the country. Though the State owned firm ECIL entered into the indigenous manufacturing of computers in 1971, it could neither fully embrace the technology and make available the systems at internationally comparable prices, nor satisfy the growing demands for the systems in the country (Subramanian, 1992). However, one point needs to be understood here is that in this new phase of globalisation we have entered globalisation too late and as junior partners where all the rules are loaded against us. Telecommunications were seen as a public utility than an instrument of economic competitiveness. Telephones were seen as a luxury rather than a necessity and the dominance of state run monopolies, the absence of shareholders pressure for efficiency, bureaucratic organizations without delegation, initiative or accountability, overstaffed structures and strong unions of workers with considerable political clout were the order of the day (Singhal and Rogers, 2001). From 1978 private sector entrepreneurs entered the computer manufacturing industry but it was the computer policy of 1984, that helped lay the foundation of IT industry (GOI, 1984). In 1985, Rajeev Gandhi's administration created the state owned Department of Telecommunication (DOT) to establish policies and grant licences for the development of this sector (Petrazzini, 1996). The DOT was assigned the task of handling imports and exports licences and ensuring that the values of socialist idealism were not abandoned. The proposed reforms were rife with opportunities for scandal and exploitation. Those with authority to grant licences, the licence permit raj and a group of private industrialists came together to oppose the opening up of the industry, slowing down any progress in the telecommunication industry.

This traditional strategy was totally reserved in 1991, when the Indian government adopted the New Economic Policy (NEP), composed of two measures stabilization and structural adjustment measures advocated by the World Bank and IMF. The former relates to the control of the fiscal balance and the balance of payments and the latter aims at improving efficiency and productivity, integrating the Indian economy with world trade and capitalist movement (Dev, 2000). In 1998, the Prime Minister constituted a national task force on IT and software. The high powered committee came out with recommendations with in ninety days of its constitution and almost all the recommendations were accepted with in a record forty five days (Chowdhry, 2002). According to this no IT company was to pay income tax on its profits for ten years Government gave land for building of IT companies. Concessions were given regarding sales tax, electricity, import duties, etc. (Chowdhry, 2002). As a result of all these efforts, India became an important player in the IT sector. The dynamics of the knowledge economy based on bargaining has made it get a large amount of concessions. The deregulation of the telecommunication had a significant impact on the internalization of the call centre services (Philips, 2002).

The software industry which started as a scratch in the early 1970's has grown at a phenomenal rate since the 1980's due to software services such as coding, customs, software development and the year 2000 Y2K work (Heeks, 1998). In the beginning much of this work was carried out at the clients facility, onsite, rather than offshore in India (Mitter, 2000). Over a period of time the scene of activity has shifted to India. In the mid 1990's Spectra mind, a Delhi based call centre and data entry site, was one of the first companies to create a base call centre site and invite transnational corporations (TNC's) to collaborate on business proposals with them, thus firmly establishing the roots of the call centres or data entry outsourcing industry (Tina Basi, 2009). Although similar developments and technological advances were evident in other Asian countries such as China, Malaysia and the Philippines (Reardon, 1991), in India the call centre industry has grown at a rapid pace over a short period of time. The IT enabled services such as Call Centers, Customer interaction, back office operations, insurance claims processing, medical transcriptions, data base management, have helped the industry move forward. In all the IT industry employed about 500,000 people in 2001 and of these 410, 000 people were in the software export sector and about 70, 000 in IT enabled sector (NASSCOM, 2000).

While it is difficult to get the exact data related to women due to the lack of gender disaggregated data in the existing literature in the National Association of Software Services Companies (NASSCOM) it is estimated that women constitute 21 per cent of the total IT force, which is higher than their participation in the National economy as a whole now at 13 per cent (NASSCOM, 2001). This is still much less than other developing countries.

ICT and Constraints to Women's Participation

Women's ability to participate in the country's ICT's growth is determined by the low status ascribed to women and girls in Indian society and the extreme poverty and poor IT infrastructure that restricts women's access to education and information technology. Women in traditional societies like India have a very low status. This is reflected in the current sex ratio. The sex ratio in 1991 was 972 females per 1000 males and in 2001 it has fallen to 927 per 1000 males (Census, 2000). The only way women can increase her status in her otherwise subordinate life is by having a son. Education, exposure and affluence have not brought in equality but only consumerism and commoditization of relationships. Female foeticide is just one of the facets of the vast anti-women behaviours spectrum in India. Hence, a traditional low value placed on girls in Indian society contributes to the family's decision to not to invest in their daughter's education. Moreover in poorer communities, girls are most valued as wage earners and so schooling is seen as a double investment. Education and literacy are low for both men and women, with women's literacy rate at only 51.4 per cent compared to men's ay 74.5 per cent. Nearly 48 per cent of the girls stay in the educational system till the secondary level only.

Since most women lack literacy and basic education, more advanced or specialized IT education is out of reach of poor women and only a reality for those middle class and elite women who can afford it. Women make up for 88 per cent of the students in arts and commerce while education and law account for 4 per cent and engineering accounts for 1 per cent of the women students (*http://projects.aed.org/techequity.India.html*). Other statistics from India also leave a lot to be desired. The female adult literacy rate is at 51 per cent and female secondary school enrolment is at 38 per cent. India ranks 112 out of 174 in internet usage. Female youth literacy is at 44 per cent and only 23 per cent of India's internet users are women (UNESCO 1999, Statistical Year Book). In many developing countries, India being no exception, women face challenges in pursuing education at all ages because of lack of time to attend school, familial and household duties and socio-cultural norms that give a low priority to education and low funds to pay for education. Hence only 1 per cent of the women enter engineering colleges. Currently IT education and training is available in engineering colleges, universities and in diploma and certificate courses at polytechnics and commercial and government training centers. Due to the heavy expenditure involved in admission to engineering colleges and in Universities most women are able to access IT training through training centers or through polytechniques. However, the quality of training varies a lot. As the job market becomes more competitive, certificates and diplomas from these institutions do not equip the job seeker with competitive skill to guarantee a job. In the absence of these and the lack of computers at home, many women rarely practice what they have learnt and further loose the chance of sharpening their skills in IT (Pande, 2012).

Historically, women's participation in the Indian labour market is grounded in the agricultural and manufacturing sectors (Banerjee, 1990; Baud, 1991). Women worked mainly in familial context and were not educated. Migration for women was very restricted and was limited to marriage and some family reason. However, in the last two decades India's service sector has replaced agriculture as the dominant sector (Chanda, 2002), with the most pronounced growth in the software service and telecommunication sector. Many women are not able to get the benefits of information technology because women's mobility is considerably more limited than men. In many areas women's mobility is further limited because social and religious customs do not permit women to travel in public without male supervision. So in many cases women may have sufficient training but do not work in the field because they lack the essential mobility to go where the jobs are. Since more women than men live in rural areas the digital divide in IT runs parallel to the rural urban divide and the gender gap. Rural areas which comprise 60 per cent of the population often lack the resources and infrastructure for IT (ITCUT, 2000). Women have a special responsibility for children and the elderly and they find it less easy than men to migrate to urban areas. This urban bias in connectivity thus deprives women more than men of the universal right to communicate (INTECH and UNIFEM).

The developing countries face three main bottle necks in engaging in an increasingly globalised, e-commerce and the information economy. These are lack of access to infrastructure, inadequacy of regulatory framework and insufficient base of human resources (Mitter, 2002). Nearly 74 per cent of India's nearly one billion people live in villages. Even after fifty two years of independence there are many villages with no basic services, 40 per cent of the 600,000 villages are without proper roads, 180,000 villages do not have a primary school with in one kilometre and 450 villages have drinking water problems. Besides these infrastructural problems the village society is often divided with complications of caste, religious bigotry and other sources of dissension. Thus the gender economy along with the political economy becomes useful when dealing with this phenomenon.

Many of the constraints which women face are due to the poverty they share with the men. Within poorest households women are poorest of the poor. Access factors also have a gender dimension to them. They have less access to money in the family than their husbands. Women here like elsewhere in the world suffer from an overload of demands on their time. Lack of contextual information, use of skills also have a gender dimension since women's isolation gives them fewer opportunities to broaden their range of comprehension (Hafkin *et al.*, 2001).

The emerging gender employment patterns in the IT sector are a cause for concern. Over the years the gender inequalities that are well established in other sectors of labour force are already being replicated in the IT sector. By and large women are well represented in desktop publishing, software programming but not in design, operating systems of computer maintenance which are still seen as male jobs. But a large number of women are concentrated in the end user, lower skilled IT jobs related to word processing and data entry. Some unskilled women have lost their livelihood as an alternative source produce goods cheaper and faster. For some women it meant a loss of rights, benefits and job security. Globalization has also led to the migration of a large number of women from developing countries (Pande, 2000). There has been a change in the nature of work also. The ICT industries require an individualisation of capacities. In the IT industry labour loses its collective identity/capacity of its workers. The position of owners, producers, managers and workers are increasingly blurred in a production system of variable geometry of team work, networking, outsourcing and subcontracting (Castells, 1997). Women who have lost formal sector manufacturing jobs frequently turn to these less desirable jobs to remain in the labour force. The work here is done at home or in small units. Here they have none of the benefits of the formal sector, like fixed contracts, maternity leave and unionizing. They suffer from lack of opportunities to work, low and discriminatory wages and exploitative conditions resulting in casualization (Pande, 2002). Teleworking is another kind of reorganization of women's work using new technology. Teleworking refers to distant working where by employers or freelance workers using telematics at a site that is geographically separated from the main site. As in IT enabled services, this also has a location set up mostly in developing countries, which is different from the main office often in developed countries. Many young girls join these and have to work throughout the night in order to keep up with the western time.

India - The Call Centre Capital of the World and Development of A.P. as Hub of ICT

India is emerging as the Call Center Capital of the world and many Multi-National Companies and Start Ups are setting up operations here. Call centres can be defined as a dedicated operation in which computer utilising employees receive in bound or make outbound telephone calls with those calls processed and controlled either by an Automatic call distribution (ACD) system or predictive dialling system (Taylor and Bain, 1999). A Call centre is sometimes defined as a telephone based shared service centre and is used for a number of customer related functions like marketing, selling, information dispensing, advise, technical support etc. Through these new technologies large amount of information can be transported at a very low cost from core office establishments to satellite or subcontracting units located around the globe. This possibility has led both private and public enterprises to externalize and decentralize non core section of their business to distant and often locations where trained human resource in IT is available at economical and greater profitable rates than at the headquarters. The outsourcing sites are often located in the developing countries such as India who have comparative advantage in providing cheap labour, who are skilled in computer use and have a good command on English language. In this context, India has become an obvious and preferred choice of several multinational IT-industries for establishing even full-fledged offices in cities such as Bangalore, Delhi, Hyderabad and Chennai. There are a number of call centres each specialising in one particular job. These include, Back Office operations which can be defined as the off site delivery of wide range of non-core service functions of any organization. Online education which can occur between two computers in the same town or between computers on two opposite sides of the earth. The Geographical Information Systems (GIS) growth driven by rapidly spreading use of mapping and spatial data technology in brand new disciplines by customer representing new industries, markets and applications. Another area which is fast developing is computer assisted animations such as graphics, animations and special effects which have become almost the sole preserve of computer systems.

Call centres can be divided into two divisions: inbound and outbound. Inbound call centers receive thousands of phone calls per day, often related to customer service. Outbound call centers may employ workers as telemarketers and sales representatives. The typical workday for a call center employee begins anywhere from four-thirty to eight pm, and usually ends somewhere between five and nine-thirty am. These abnormal office hours are implemented in order to accommodate clients based in the United States. 4:30 pm in India translates to 7 am on the east coast of the United States, while 8 pm in Hitec City translates to 7:30 am on the United States' west coast. Therefore Indian employees are trading their days for their nights.

The international Data Corporation estimated that spending on IT outsourcing has reached $56 billion in 2000 and the revenues for 2008 are estimated to be over USD 50 billion. From 1998 to 1999, the IT enabled services employed 23,000 people and in 2001-02 this number rose to 106,000 and in 2003-04 it was 245,000 (NASSCOM, www.nasscom.in). India has become an obvious choice of the industry because it provides a large pool of English speaking people. In some ways the spread of IT enabled services has been immensely beneficial to both women and men especially those who have limited skills and lack the resources to invest in higher education. Minimum wages in India are just over Rs. 1000 per month, about 20$, but in the IT enabled industry it could range from Rs 5000, 102$ to Rs. 15000 about 306$ (Kelkar, 2002).

With the rise of IT sector in India, the state of Andhra Pradesh has also emerged as the hub of ICT activities. It has adopted various progressive steps towards this purpose. The training of engineers has been stepped up and the number of engineering colleges went up from 35 in 1987 to 215 in 2002. Besides this it also started MCA and BCA programmes to provide skills in software. International Institute of Information Technology, IIIT was started by getting a number of private companies to come together to found a University. Cyber Towers in the Hi-Tech-city, Hyderabad, provides 8,50,000 sq.ft. of office space for software companies. The Government of Andhra Pradesh also provided many incentives to IT companies. For instance if any IT company which would invest more than Rs. 500 million would get land at Rs. 5 million per acre and a rebate of Rs. 20, 000 per job created from the total amount payable for land. It is promised uninterrupted power supply and exempted from statutory power cuts and guaranteed round the clock maintenance (Chowdhry, 2002). Besides this the Government also amended the A.P. Shops and Establishment Act that regulates employment conditions of workers and young persons. Section 21 and 23 of the Act have been amended to employ women and young persons between 18 to 20 years. Employers now have a flexible timings for work and can work for more than eight hours and they can operate 24 hours a day for all 365 days a year (Chowdhry, 2002).

The employment of women and girls has grown enormously in these call centres. Women usually predominate in these services requiring rather routine, low level or limited technical training. These include activities such as customer call centre, data entry, processing transcripts, service claims processing and remote sectarian services. It is very difficult for these women to move into higher skilled better paid jobs such as software development programming or geographical information system analysis. There is no scope for up gradation of skills and women continue to work in the same kinds of jobs with only degrees of variations. There is a clear-cut hierarchy in these call centres with finance, marketing, credit cards, insurance, stock markets and bonds at a higher level sales and marketing are next in status and is more output oriented. The third level is of customer care which is more quality oriented. By and large this job is of contractual nature and due to this it precludes many of the essential security mechanisms (Pande, 2005).

In some ways the spread of IT enabled services has been immensely beneficial to both women and men especially those who have adequate skills but lack the resources to invest in higher education (Pande, 2006). In these offices job responsibilities have been restructured into two main components, one "administration support" that deals with decision making and another "technical support" that is mainly responsible for execution of tasks. The division of job responsibilities has an infrastructural support of advance tools in IT technology such as high speed internet, global connectivity to work-stations, and high volume communications networks. In general it has made possible to spatially separate routine back office tasks from non-routine works (Greenbaum, 1995).

The Realities of Call Center Employment: A Case Study

We conducted a study in some of the IT enabled services in the twin cities of Hyderabad and Secunderabad in Andhra Pradesh, with a sample of 75 women. It was very difficult for us to talk to these women in the companies where they were working such as G. Capitals, Oracle, Microsoft, Deloitte and hence we got the address of these women and spoke to them at home. We also interviewed girls working in some medicals transcription centres. These interviews were carried out from 1st July, 2011 to 30th December, 2012. We did not have any structured questionnaire but interviewed the women with certain broad based questions in mind. We had an in depth discussion on work, family, health and future aspirations. Most of the girls whom we interviewed were in the age group 23 to 29 yrs. The girls in the companies were very hesitant to be interviewed and they did not allow us to tape their interviews. However, it was through a network of friends that we were able to speak to these girls outside their working hours.

In our sample there were 30 graduates, 20 post graduates, 15 engineering degree holders and 10, besides having a graduate degree were computer diploma holders. Of these 75 girls, 40 had been working for two years and 30 for less than a year, and 5, for less than two months. In our sample of 75, only 20 were married out of which 11 had children. All the girls whom we interviewed were smartly clad and some of them did wear the traditional Indian dress, *Salwar Kameej,* but a large number of them were in western attires. Irrespective of the attire all of them appeared to be go getters, slick talkers with an extraordinary flare for communications with impeccable mannerism and vibrant dynamism. The ideal software worker is also an ideal product who is as much a product of the marketing packaged image where we see the ever fresh young girls in the BPO's and the KPO's who hide their older sorority who have quit and thereby cause large scale attrition.

All the girls in our sample had undergone the voice modulation test for removing traces of regional slang. We also visited two schools which provide training in fluency in English, pronunciation, diction correction and appropriate usages in spoken English suitable to the clientele and had a look at their training schedule and attended some of the classes which they conducted. A major part of the training consists of neutralising the accent. Initially all of them had difficulty in understanding the English language spoken by the Americans but they picked it up soon and today all of them are experts in understanding the American accent. During training, employees are put through a two to four week training course, in which they attempt to minimize their Indian accents, and become familiar with certain phrases. Voice recordings are made at the beginning of the course and then throughout the programme, so that students can observe their progress. " *We were made to speak on the first day and the same sentences were recorded by a professional person and when the tape was played to us, I could not but laugh at how I was speaking English and all the while I thought I was much better off than most of the girls because I could speak English fluently*" (Reena, personal interview, June, 2011). The classes focus pronunciation of vowel sounds and certain consonants. The pronunciation of the "v" and "w" is highly stressed. Common phrases and slang are also addressed. "*It's the little things,*" commented Romila, "*Instead of saying, 'May I know your name, like I usually do here, 'I say may I have your name.'* (personal interview, June, 2011). These subtle changes make communication easier and more civil. "*Mam, they have improved my English and when I listen to my earlier tapes and accent I feel very ashamed that this is how I used to speak earlier*", said Arunmai (Personal interview, June, 2011). The change in their accent gave these girls a sense of becoming more and more modern.

Despite the lengths that call centre employees go through to assure effective communication with their clientele, negative incidents still happen: "*Sometimes when people hear someone Asian is calling, they just hang up the phone.*" (personal interview, Malvika, January, 2012). The same respondent also discussed how her feelings were often hurt by derogatory comments that she received over the phone. Students interviewed in the University of Hyderabad, in the student film, *Day for Night*, also replicated these sentiments: "*People would say to me, can't you speak English?!*" (personal interview, February, 2012). Furthermore, the E- Commerce and Development Report, 2002, recognized that call centre employees were required to civilly deal with callers, "...who tend to be abusive or even hysterical." However, it is standard for companies to have a hierarchy of resources for dealing with irrational customers. A call center employee can easily transfer a call to a supervisor, if necessary. On the other hand, there is evidence that many calls between Indians and American are quite successful. One respondent commented that one of her favourite parts of the job was talking with Americans. Often times, companies receive compliments in the form of letters and phone calls regarding the excellent performance of their call center employees. Therefore, call center employees are at a risk of being either berated or praised by a client.

Aside from client interaction, call center employees are also subjected to various physical and emotional side effects. Due to the abnormal working hours, employee's internal biological clocks have to be reworked. The body's normal cycles of sleep are completely disrupted, causing sleep deprivation, fatigue and irritability, along with vomiting sensations and severe headaches. The persistence of these symptoms depends on the individual, ranging from three weeks to several months. However, once adjusted, employees are quite comfortable with the hours, and enjoy not having to wake up early in the morning for work. As mentioned above, a lot of employees are younger and fresh out of university, which likely contributes to their resilience in dealing with the nonstandard office hours. Aside from the physical, call center office hours also affect social interaction. Call center employees are disconnected from their families and friends, and in some cases go for days without seeing the people they live with because of the drastic differences in their schedules. Kavita, a young woman working at an outbound sales call center, has not seen her mother in three days, despite living in the same house! (personal interview, 2004). The lack of outside interaction turns social focus inwards; call center employees often spend their time outside of work with other call center employees. In this sense, call center colleagues become like extended families, and significant socializing goes on within call center walls.

In addition to language classes, employees also take up stage names. Using a false (Western) name negates the Hitec employee from having to take time to spell out a foreign name to his/her client. The company that an employee works for determines stage name guidelines; some require employees to use names completely different from their own, while other companies promote the use of similar sounding names. Most of these girls are given a professional name so that the clients in the west are able to get their names easily. Hence, Bhagyalaxmi becomes Betty, Aishwarya becomes Alice, Srilakshmi becomes Sherry, Devyani becomes Devon, Varalakshmi becomes Veronica, Divya Sai becomes Diane; Shashi becomes Susan and so on. Thomas Macaulay envisaged in his famous minute on Education of 1835, "a class of persons, Indian in blood and colour, but English in taste, in opinions, in morals and in intellect…." the call centre creates new forms of social divisions, separating these reconstructed young adults from the rest of the society. It reinforces social gulfs, alienating people from their traditions, without offering them any place in the values they have to simulate in order to ease the lives of distant consumers they will never meet (Seabrook, 2003). Of course, the young people will not become babus in the administration of empire but, they have their function. Their conspicuous un-Indian lifestyle signals to a generation that they, too, can be liberated into the have-a-nice-day culture of global fantasy. Though these girls are Veronica by night they continue to be Varlakshmi by day.

The call centres have created a greater sense of western identity. These women dress differently, talk differently and look more westernised and sexually liberated. However, this does not conflict with their Indian identity and these women regard their western name as a requirement of the job. In spite of the arguments that the call centre girls are becoming more westernised many of these women have a sense of Indianness heightened by their experience of working in a call centre. *We are doing a job and are better off than many other girls of our age. We know so much about different countries and their culture. Just because we are working at night does not make us more western or loose women* (Riya, personal interview, March, 2012). Many of these girls move out from their home towns due to limited opportunities in their home towns. The process of urbanisation and the large number of out station employees meant that women in these call centres increasingly identify themselves regionally. We found the Oriya girls networking with each other and so were the Bengali girls. They moved out together in their own groups and preferred staying with the person from their own region. Call centre have become like internet kiosks and cyber café have opened up on every corner giving Indians access to global communications (Das, 2000). Although the call centres enhance a perceived westernization of Indian culture, European, American, cultural influences can be found in many urban centres, cinemas, multiplexes all over India in metropolitan areas. Indian culture is certainly not a concrete or unmovable culture, but is very much influence by television, movies and the changing pace of industrialisation and urbanisation. The belief that the call centres are solely responsible for cultural changes is some what naïve for the globalizing process can be found in shopping centres, cinemas, multiplexes, bars, restaurants and universities.

Many of the girls that we spoke to told us that it was the people who had a perception of their westernness and they were very much Indian to the core but more modern and in control. They also knew that many people had a very negative perception about their working at night. This negative perception conflicts with their own identity constructed as independent, liberated and modern women providing them agency. "*Initially my parents were not very keen that I work in the call centre but then my father came to Hyderabad and met the other Bihari girls, came to know about their backgrounds and now they are happy with my work* (interview, Geeta, May, 2012). *It does not trouble me what people think of me and my work. I have friends working with me both boys and girls and if they understand my work, nothing matters as to what others think. My family is broad minded and they are not bothered as long as I am safe.* (Personal interview Vanaja, May, 2012). One has to view globalization as a series of scrapes and global cultural flows (Appadorai, 1996). Culture is something which is not fixed but which is constituted and reconstituted over time.

I am very happy with my work for I have enough money to lead a good life, I can go to movies, discos and even save money for my marriage (Personal interview, Meena, June, 2012). Although most employees are satisfied with their current working conditions, another phenomenon of call centres seems to be high turnover rate, due to the unique nature of the work. One complaint is that the actual tasks completed at a call center are decidedly repetitive; employees become apathetic and bored, and often shift companies. Instead of working for an outbound telemarketing firm, an employee might shift to an inbound insurance claims processor. If boredom isn't the trigger, too much stress may be the catalyst for a call centre employee quitting his/her job. There is a prospect of a 'burn out' syndrome, where employees are simply tired of the negative physical and emotional effects that working in a call centre can sometimes create. Danger also lies in a call centre's limited, career promotion prospects. Respondents alluded to the fact that working in a call centre limits an employee's experience in other fields. Effectively, working in a call centre only really prepares you to work in *other* call centers, not necessarily other fields like banking or development. A bridge has been established between former call centre customer service representatives and human resource/management positions, but that doesn't ensure a seamless transition. To lower the attrition rates many companies try to build social bonds between employees and hold social gatherings and take their employees out of station on a holiday during Christmas or New Year.

As all the customers are at the other side of the globe, all these girls work at night. Initially the girls found the jobs to be interesting but soon it became monotonous. They work in shifts of eight hours, and work varies with period of lull and high stress. All the women we spoke to acknowledged that they worked under sever stress levels. Most of the Call centres have systems of monthly performance appraisals or ratings. Work repetitiveness, stress and lack of career prospects were the reasons for leaving the Call centre job. Many girls complaint about the clients using very foul language and being short tempered. "*I used to enjoy watching American films on TV, but now when I see that channel it reminds me about my clients and change channel*" said Rajshree. The first things these girls are told is that they should never disclose to the client that they are from a faraway country (India) and city like Hyderabad. They have to know the weather forecast of the country with which they are dealing mostly, USA. The girls have to memorise the details about the local football team so that no, "Dead air" occurs. "*Often when it is pitch dark outside, we end our conversation by saying, What a lovely sunny day, enjoy the weather and have a good day Mam*" said Roopa (Personal interview, May, 2011) " *We have to reinvent ourselves and no matter how big a festival we may be having in India like Deepawali, Dushera, Holi or Ugadii, it does not have any significance and we work at nights but now we are more involved with Halloween, Christmas and Easter*", said Neena (Personal interview, June, 2011). They have to repeat thank you often, which does not come so easily in Indian culture. "*This job is very tough and my friends think that I am lucky to have got this job but I envy my friends who are sleeping cosily at home, while I work at night*" said Devyani (Personal interview, July, 2012). " *To me this is only a stop gap arrangement till I get a decent job during the day time*" said Nandita. Even Ranjita told us, "*I do not think one can survive in this job for more than two to three years. I now feel that I do not mind a less pay but working hours during the day is the first and foremost condition in my job priority*" (Personal interview, July, 2012). All these girls knew very well that they had to bear all this hardship and manage their grave yard jobs as they call them, because the pay packet, which they got at the end, was much more than they could ever get with their qualifications. "*I have often thought of leaving this job, but I have to stick around for the sake of the money*", said Parveena (Personal interview, July, 2012). To break the monotony of work these girls are made to do the mandatory exercise for ten minutes, which the "management insists is for their wellbeing". Sitting in one chair for eight hours is very taxing and painful. Although the chairs are said to be designed for relieving stress, end up with minor back aches, which at times turns out to be severe. Now after having worked for a long time they are all accustomed to getting up at night but they never seem to get enough of sleep.

"*Sleeping throughout the day can never compensate for the night sleep and so I sleep in bouts and every little sound, and there are plenty in the house during the day that disturbs me*", says Nageswari (Personal interview, 18th November, 2011). *There are so many disturbances in the morning though we have rented a room. The maid comes to the next door house and there is a lot of noise, there is their dog which barks constantly and I find a lot of loud noises says Neena,* (, Personal interview, 15th August, 2012). All the girls complain of the different sounds during the day, the vendor shouting, people ringing the call bell, neighbours dropping in for a chat and all these irritate them. They tell us that they know it is not possible to have pin drop silence like at night time and hence they never seem to get enough sleep. The management tries to ease these by providing food. They also have a lot of health foods and provide a lot of salads to keep their workers healthy. Many of these girls also complain of gastrointestinal problems which are inevitable for those working at nights as the body is put under chronic stress. As most of these jobs are at night which means readjusting the biological clock to a different rhythm, the basic socio-health hazard remains. At the end of the day, or night, the girls are harrowed with stress, sleep and tired of putting up that plastic smile but subconsciously accept that they are paid for it.

All the managers to whom we spoke to justified feminine traits like patience, soft and sweeter voice for the preponderance of women over men. None of them recognised that the reason women were in predominant position was due to the nature of the job which was very monotonous. "*I think by nature women are better behaved and gentle to deal with strangers. So they are the ideal choice for this job*" said Mr. Devashish the manager in one of the Call centres. "*It is not that we do not get boys but we prefer a girl over a boy because they have more patience and can solve problems with out getting frustrated or showing their impertinence*" said Mr. Deewakar another manager. Many female executives also agreed with this. But the reality is that this industry thrives on the availability of cheap and flexible labour of women.

Many girls that we spoke to welcome these jobs due to secure physical environment and free transportation provided. Many of these girls are picked up from their homes and dropped back after work and this seems to be the biggest advantage of doing this job as far as many parents were concerned. We found that besides the lack of jobs in other sectors, this job is favoured due to the media hype and the consequent eulogising social status as being part of a most happening industry. But, the women whom we interviewed hardly relate to this industry in these terms. They did not see their lives as glamorous or being part of the most happening industry.

This job definitely gave a lot of mobility to unmarried girls and they did not have the burden of house keeping. About 38 girls in our sample had migrated from Bangalore, Calcutta and small towns and lived in paying accommodations or hostels. This was something unheard of five to ten years earlier and parents would never send their daughters to a new place to live and work. "*I like this job because it gives me the financial independence and to do as I please. I cannot think of such a salary with the present job scenario in India*", said Ritika (Personal interview, June, 2011). To Sushmita, this job gave the freedom to move away from the restrictive environment of her home in a small town of Orissa and she enjoys much more personal freedom in a big city like Hyderabad (Personal interview, July, 2012). It was interesting to note that in spite of all the benefits which these girls saw in working in a Call centre, none of the unmarried girls look upon this as a career prospect. Many of the girls looked upon this as a stop gap arrangement before marriage.

Of the twenty married girls in our sample eleven had children. They had help or husbands who took care of the children. Five of these girls whose children were going to school, were seriously contemplating of leaving this job after six months. All these women stated that they were not able to devote time to their homes and children. They had to cut down a lot on social life, social functions and often it was their husbands who attended these alone along with the children. "*I come back home at seven in the morning and I want to just hit the bed and go to sleep but then I have to dress up my child and get him ready with his tiffin and I can only go to sleep once he leaves for school around eight thirty, tells us Savita*" (Personal interview, 18th August, 2012). By and large the married women felt disadvantaged because of their socially defined responsibilities of child care and housework. They felt that there was no change in their social status. They felt that they were doing the job as they needed the money. Since many of these women were very busy with their work they had little time to take charge of the house and hence decision making was primarily taken by men or elders in the family. We found that married women were still bound by domestic and child care responsibility, restricting their mobility for career enhancement. Married women felt that keeping the house and looking after the children was their main responsibility. On holidays the women took charge of household works completely. The women we met were trying very hard to be efficient workers at office and at the same time be perfect housewives at home. Married women, though aware of the changes occurring with their working in IT, did not feel competent enough to challenge embedded patriarchal relations and the existing structural inequalities that go with it. These women are caught between the images of a traditional Indian women and the new liberated women while such symbol enhance the image of Indian economic liberalisation, they do not necessarily afford women liberalization from the patriarchal discourse. These women are caught between the images of traditional Indian women and the new liberated women. While such symbols enhance the image of Indian economic liberalisation they do not necessarily afford women liberalisation from a patriarchal discourse.

Conclusions

While Information and Communication Technology has become a potent force for transforming social, economic and political life in the globalise world, the gendered division of labour is already emerging. In ICT, women are excluded by and large at the higher level and they are confined to the lower level in IT enabled services or call centres. The gender gap- especially the gap between men and women and how they benefit from Information technology has widened because women are less likely than men to receive technical education or be employed in technology intensive work. This is relatively true in rural areas among poorer population where women are already disadvantaged by their relative lack of education. Information technology is both an enabling and limiting factor. While globalization has brought in many new opportunities and increase in income it has not really changed the gender division of labor in the workplace and at home or subverted patriarchal norms. While there is a perception of very westernized liberated women, these women have a distinct Indian identity that is being continuously redefined and altered. Ideas of domesticity and traditional subservience are invoked in the home while cementing the same with the role of a breadwinner. Similarly the idea of a feminized labour force is also created, of a person who does not think beyond the glass ceiling. Call centers present a very complex picture in terms of empowerment, agency and impact on the women working there.

References

1. Appadorai, A. 1996, *Modernity at large, cultural dimensions of Globalization,* University of Minnesota, Minneapolis and London.
2. Banerjee N. (ed), 1990, *Indian women in a changing Industrial society,* Sage Publications, New Delhi.
3. Baud, IAS, 1990, In all its manifestations : The impact of changing technology on the gender division of labour, in Banerjee N. (ed), 1990, *Indian women in a changing Industrial society*, Sage Publications, New Delhi. Pp. 33-132.
4. Barber B, (1993, 2004), Jihad vs Mc World, in, Lechner F and Boli J. (eds), *The Globalization Reader*, 2nd edition, Blackwell publishing, Oxford, pp. 29-35.
5. Boserup, E. 1970, *Women's role in economic development,* Aleen and Unwin, London.
6. Butler J. 1990, *Gender Trouble: Feminism and the subversion of identity,* Routledge, New York.

7. Chanda, R. 2002, *Globalization of services : India's opportunities and constraints,* Oxford University Press, Oxford.

8. Chowdhry, T.H. 2002, Information Technology for development, Necessary conditions, *Economic and Political Weekly,* September, 21st.

9. Castells, 1996, *The rise of the network society,* The information age: Economy, society and Culture, Vol. 1, Oxford University Press.

10. Castells, 1997, *The power of identity:* The Information age, Economy, society and culture, Vol. 2, Oxford University Press.

11. *Census, 2000,* Government of India and Registrar General of India, Reported in Health Statistics of India (1985), Central Bureau of Health Intelligence Directorate General of Health and Family Welfare.

12. Das, G, 2000, *India Unbound: From independence to Global information Age,* Penguin Books, New Delhi.

13. David, O Stephens, 1999,The globalization of information technology in multinational corporations, *Information Management Journal,* July, 99, Vol. 33, Issue 3.

14. Dev, S. Mahendr (2000), Economic liberalization and employment in South Asia, *Economic and Political Weekly,* 8th Jan., pp.40-51.

15. E- Commerce and development Report, 2002, http://unctad.org/en/Docs/ecdr2002_en.pdf, Accessed, 17th March, 2013.

16. Fatma, Alloo, 1998, *Women in the Digital Age- Using Communication Technology for Empowerment,* A practical Handbook, UNESCO, Rome.

17. Featherstone M. 1995, *Undoing culture: Globalization, Postmodernism and identity,* Sage Publications, London.

18. Freeman Carla, 2000, *High Tech and High Heels in the Global economy; women's work and pink collar identities in the Caribbeans,* Duke University Press, Durham, N.C.

19. Greenbaum, Joan, 1995, *Windows on the workplace: computers, jobs and the organization of office work in the late twentieth century,* Monthly review Press, New York.

20. GOI, 1984, *The New Computer Policy,* The Department of Electronics, Government of India, New Delhi.

21. Hardt M. and Negri A. 2000, *Empire,* Harvard University Press, Cambridge, London.

22. Hafkin, Nancy and Nancy Taggart, 2001, *Gender, Information Technology and developing countries,* An analytic study, Learnlink Project, Washington D.C.

23. Harindranath, G. 1999, Information technology in India: Production, diffusion and policy trends in Felix B. Tan, P. Scott Corbett, Yuk Yong Wong (ed), *Information Technology Diffusion in the Asia Pacific, Perspectives on Policy, electronic, commerce and education,* Idea Group of Publishing, Hershey, USA.

24. Heeks, R.B. (1996), *India's software Industry: State Policy, Liberalization and Industrial Development,* Sage, New Delhi.
25. Huntington, S.P. (1993, 2004), The clash of Civilization in, Lechner F and Boli J. (eds), *The Globalization Reader*, 2nd edition, Blackwell publishing, Oxford, pp. 29-55.
26. ILO, 2000, *World Employment Report,* Geneva.
27. ITCUT, 2000, Report of International Telecommunication Union Task Force on Gender issues.
28. INTECH and UNIFEM, Gender Telecommunication: An agenda for Policy, http://www.itu.int/ITU-D-Gender
29. Kraemer, K, 1994, *IT and economic development: Lessons from the Asia Pacific Region,* PICT Policy Research Paper No. 26, Programme on Information and Communication Technologies (PICT), Uxbridge, Brunel University.
30. Kelkar, Govind, Girija Shreshtra and Veena N. 2002, It industry and women's agency, Explorations in Bangalore and Delhi, India, *Gender Technology and Development,* 6 (1), Sage, New Delhi.
31. Mitra, Sugrata, 2001, Paper in USAID Conference on *New Technologies for development and disaster relief,* January, 2001, Washington D.C.
32. Mitter, Swasti and Sheila, Rowbotham, (eds), 1995, *Women encounter technology, Changing patterns of Employment in the third world,* Routledge, London.
33. Mitter, Swasti, 1999, Globalization, Technological changes and the search for a new paradigm for women's work, in Gender, Technology and Development, Special issue on New Technologies and Women's employment in Asia, 3, (1).
34. Mitter, Swati, 2000, *Economic and Political Weekly*, Vol. XXXV, No. 20.
35. Mitter, Swasti, 2002, Background paper prepared for UNCTAD experts meeting on, *Electronic Commerce Strategies,* 10th to 12th July.
36. Nash J. and M.J. Fernandez, Kelly (ed), 1983, *Women, men and the International division of labor,* Sunny Press, Albany, New York.
37. NASSCOM, 2000, *IT Enabled services: Background and Reference Resources,* NASSCOM, International Youth Centre, New Delhi.
38. OECD, 1987, *Information technology and Economic Prospects,* Information, Computer and Communications Policy series, ICCP No. 12, Paris.
39. Pande Rekha, 2012, Gender Gaps and Information and Communication Technology.
40. A case study of India, in Pande Rekha and Theo P. van der Weide (ed.) 2012, *Globalization, Technology Diffusion and Gender Disparity: Social Impacts of ICTs,* Information Science Reference, IGI Global, Hershey USA, pp. 277-291.
41. Pande, Rekha, 2006, Digital divide, gender and the India experience in IT, Vol. 1, *Encyclopaedia of Gender and Information Technology,* (ed) Eileen M. Trauth, USA, Pennsylvania State University, pp. 191- 199.

42. Pande, Rekha, 2005, Looking at Information Technology from a gender perspective: The call centers in India, *Asian Journal of Women's Studies,* Volume 11, No.1, pp.58-82.

43. Pande, Rekha 2002, Women's work in the unorganized sectors in India, in Meiko Sugiyama (ed) A *cross Cultural study on the life of women*, Women's Studies Series, 22, Japan, Tokyo Women's Christian University, pp 58-87.

44. Pande, Rekha, 2001, The Social costs of Globalization: Restructuring Developing World Economies, *Journal of Asian Women's Studies,* Vol. 10, December, Japan, Kitakyushu Forum. pp. 1-14.

45. Pande, Rekha, 2000, Globalization and women in the agricultural sector, *International Feminist Journal of Politics,* Rutledge, U.K. Vol. 2, No. 1. pp. 409-412.

46. Personal Interviews with Call center employees, G. Capitals, Deloitte's. These interviews were carried out from 1st July, 2011 to 30th December, 2012.

47. Petrazzini B.A. 1996, Telecommunication policy in India: The political underpinnings of reform, *Telecommunications Policy,* Vol. 20, N0. 1, pp. 39-51.

48. Philips A. 2002, What it was like, What happened and what's it like now: development in Telecommunications over recent decades, *Journal of Regulatory Economics,* Vol. 21, No. 1, pp. 57-78.

49. Reardon, G. 1999, *Globalization, Technological change and women workers in Asia,*Maastricht, Netherlands, UNU/INTECH.

50. Seabrook, J, 2003, Progress on hold, *Guardian,* 24th October. http://www.guardian.co.uk/world/2003/oct/24/india.comment, accessed 18th March, 2013.

51. Singhal A. and Rogers E.M. 2001, *India's communication Revolution: From Bullock carts to Cyber marts,* Sage Publications, New Delhi.

52. Subramanian, C.R. 1992, *India and the Computer: A Study of Planned development,* Oxford University Press, New Delhi.

53. Taylor P and Bain P, 1999, An assembly line in the head; Work and Employee relations in a call centre, *Industrial relations journal,* Vol. 30, No. 2, pp. 101-17.

54. Tina Basi, J.K. 2009, *Women, identity and India's call Centre Industry,* Routledge.

55. UNO, 2000, *Okinawa Summit Report: Group of eight Digital Opportunities,* Task Force, Okinawa.

56. www.idc.com, accessed on 16th March, 2013.

57. www.nasscom.org, accessed on 16th March, 2013.

58. Ward, K. 1984, *Women in the world system- its impact on status and fertility,* Prager, New York.

59. Wajcman, Judy, 1991, Streetan, *Feminism confronts technology,* Oxford Policy Press.

Chapter 19

High Performance Computing in Materials Science Research: Are There Women?

Rapela Regina Maphanga

Associate Professor, Materials Modelling Centre,
University of Limpopo, Sovenga 0727, South Africa
E-mail: maphangarr@yahoo.com, rapela.maphanga@ul.ac.za

ABSTRACT

This paper gives a brief overview on representation and visibility of South African women in high performance computing research. The overview has provided insights into the importance of high performance computing in solving materials science research and its contribution to the well-being of global community. Socially constructed gender roles and relationships play a key role in determining the capacity of women and men to participate on equal terms in the information society. Despite the fact that women comprise more than half of the world's population, they are under represented as a group in information communication and technology, particularly in high performance computing (HCP). The paper spells out role that women can play in high performance computing when the gender gap is narrowed. Women, for example, are less likely to own high performance computers and clusters for their research activities. In addition, women's access to and use of information and communication technology (ICT) is constrained by factors that go beyond issues of technological infrastructure.

Keywords: *High performance computing, Materials science, Women in science, Information communication and technology, Gender gap, Gender equity.*

Introduction

There are more females at universities as compared to males globally. More institutions of higher learning except engineering schools have a ratio of 60:40 females to males. This include Harvard University which has been a male bastion has begun to tilt toward women (52 per cent of their students were females). Lewin (2006) reported that more women complete their undergraduate degrees as compared to men. The current South African status is that, women are in the majority in terms of enrolments and graduations and the number of female graduates is rising annually. Despite women being in the majority at Universities in Europe, USA and Japan, men in all areas of workforce do get better pay and promotions because they tend to work longer hours and have fewer career interruptions than women. Women get interruptions to bear children and carry the most responsibilities in raising them (Lewin, 2006).

Throughout history, women worldwide have occupied lower positions in society. Gender imbalances are attributed to socialization and extra-organizational gender rules which ascribe domestic labour to women. Across the globe, women's under-representation in academia has been documented, for example in: the United Kingdom (Universities UK, as cited by Sanderson, 2002), Korea (Kim *et al.*, 2010) Portuguese (Carvalho and Santiago, 2008 and 2010), Australia (Gardiner *et al.*, 2007), and USA (Hult *et al.*, 2005; Mayer and Tikka, 2008; and Townsley and Broadfoot, 2008). On the contrary, no reliable statistics on women representation in ICT in different countries is available.

Worldwide women experience conflict between their domestic and professional roles (Lewin, 2006). Women are late starters as well as late achievers because they have to interrupt their career or work to care for their families. Women are supposed to choose between dedicating their time to family or building their career in any field; then start family later in life. In addition, previous studies reported that some married women miss the opportunity to be promoted due to the fragmented educational and career patterns as a result of their husband's career movements (Theron, 2002). The lists below are some of the obstacles of women's career development:

- Women are under valued, under recognized and under rewarded as compared to men.
- Women are often recruited with less tenure than their male counterpart.
- Those who are in leadership are isolated and do not receive mentorship, collaboration, because only fewer women in such positions and most of the networks are gender based and occur under informal situations.

In the past two decades major advances on protecting women's rights and gender equity perspectives were made. Gender equity and economic structures are closely linked. Gender equity has been promoted by international organizations such as the UNESCO and the World Bank. A literature with theoretical origins in neo-classical economies has focussed on the male-female ratio equity in household decision-making and allocation of resources, and on the economic and social benefits of education for girls and women as a form of human capital investment (World Bank, 1994).

Advocacy for gender issues in ICT gained its first international foothold at the 4th World Conference on Women in Beijing in 1995. Since then, gender issues in ICT policy have been on the international scene. In the social fields such as health, education, economics, agriculture and rural development, gender issues are considered to the most of their projects. However, the presence of gender issues extends to ICT. Women have less access to resources such as infrastructure, finance, technology and education needed to support their active engagement in ICT. As a result, their presence in ICT workforce, entrepreneurship and research is lower than men's. Women living in developing countries with low social status face life with challenges and disadvantages that are difficult to overcome. These challenges persist as women strive to access and use ICT to better their lives. Countries such as South Africa try to put supportive policies in place to cater for gender issues. The SA gender policy was adopted by cabinet in December 2000. The policy framework outlines South Africa's vision for gender equality and sets out how it intends realizing its gender goals. It establishes guidelines for the country to remedy the historical legacy of racism and sexism by recommending an institutional framework that facilitates equal access to goods and services for both women and men. The South African gender policy outlines three main areas of intervention that government should make to mainstream gender and interventions, namely:

- Promoting women's empowerment and gender equality in their service.
- Raising public awareness about gender in their interactions with clients and stakeholders in the private and public sectors.
- Promoting women's empowerment and gender equality in their employment policies and practices.

Method

The methodology employed in this study included documentary analysis from the Department of Science and Technology review report on the status of women in SET, Meraka Institute of Council for Science and Industrial Research of South Africa. The statistics gave a contextual framework for an understanding on current status of women in ICT in SA. Documents such as those used in this study are deemed attractive to social scientists for the purpose of analysis as they are perceived to be authorities, credible objective and factual (Denscombe, 1998).

Science and Technology

Modern developments in agriculture, communications, health, industry, and materials are driven by science and technology (S&T). S&T is one of the most productive systems created by man, in the sense of resulting in unprecedented progress. The societies that have embraced S&T as the major engine for growth and incorporated it into their own economies, cultures and traditions have a better life. Nations that are better positioned with respect to S&T are wealthier and enjoy a better quality of life. Most countries which do not embrace science and technology would remain stagnant in their social development and deteriorate in their economic development. It is a well-known fact that one of the major global challenges is to promote economic growth while at the same time providing for social development as a means of addressing the needs for the poor.

The growth and development of every nation in this era is directly linked to the development of science and technology in that particular nation. Proper development of these areas of human endeavour is, without doubt, a priority for faster delivery of services to every society. A better placed nation with respect to science and technology, the greater is its economic growth and development, with consequent of higher level national wealth and a better quality of life. Science and technology has a profound effect on nation's ability to compete in the global markets. It contributes to employment, health and prosperity worldwide. There are key elements of a nation with a strong science and technology base, this include: excellent educational system, determined investment in research and development, inspired mentoring by champions, visionary leadership and public awareness of and support for the science and technology.

South Africa currently faces several constraints in its attempts to move up the science and technology innovation value chain. The challenges include: the partial impact of new industry policies, poor performance against international benchmarks, a weakening science base and the decline in the national science and innovation system. The key inputs in the process of moving up the value chain include: enterprise restructuring, government science and technology policies, and the ongoing commercial application of science and technology innovation.

In recent years, science and technology experts, professionals and scholars (including women) have been seen appearing in parliament, cabinets, national and international unions. They are articulating visions, mission statements, action and business plans and timelines for science and technology, and producing policies and making significant decisions for global S&T.

Women in Science, Engineering and Technology

Gender equity issues represent major challenges for all science, engineering and technology fields. Women are historically under-represented in these fields, and their sense of integration in the community is often low. Despite tremendous efforts to give women greater access to education in science, engineering and technology in most countries, there is still significantly under-representation of women in some science degree programmes, especially in engineering, physics, computer science and ICT. Also, even though the access to science and technology education is improved, however, number of women in the workforce is not increased. In fact, in some countries including the USA, the number of women in the science and technology workforce is declining. The Science Council of Japan, in its report *Japan Vision 2050,* expressed its concern about male domination in science and technology and recommended that the issue should be addressed (Science Council of Japan, 2005). This is because that the Japanese women employees in S&T are confined to low levels of the occupational hierarchy.

Gender equity is one of the core ideals preserved in the Bill of Rights of the South African Constitution (1993). The South African government has adopted the approach of 'gender mainstreaming' in order to address and respond to gender inequalities. Thus, the government has established 'gender desks' or 'gender focal points' within each government department, whose responsibility is to ensure the integration of a gender perspective into all policy and implementation activities. The government has also developed policy and legislation that tagged women as a specific group *e.g.* Employment Equity legislation.

In her address during the Third World Organisation for Women in Science (TWOWS) Conference in 2010, the minister of science and technology Honourable Minister Naledi Pandor said:

"Women are still less represented in the fields of science and technology. Women are still less represented in the top research managerial positions. Women are still under-represented in science, technology, and innovation policymaking. The challenge for Africa and the developing world is to ensure that the gender imbalance in the practice of science, technology and innovation activities is addressed."

She stressed the importance of science, technology and innovation (STI) for socio-economic development in both the developed and developing world. The involvement of women in STI activities is critical in ensuring that the full diversity of a nation is utilized in providing expertise and in contributing to the development of nations.

It is evident that numbers of women in science, engineering and technology (SET) are alarming low, even in the world's leading economies and are declining in many countries, including in the USA. South Africa is not an exception, even though the South African women have more opportunities available to them than ever before, their participation in SET workforce remains low. The South African policy framework addresses SET for women and look at guidelines for the development of SET products and services, which will benefit women living in poverty and rural areas. Although the policy does not specifically address for ICTs, the broader SET framework set the direction for any future strategies addressing gender equity in ICTs. A supporting research was carried out in 2004/5 to provide the knowledge base for the formulation of gender equity policy for science, engineering and technology. There are various publications produced by the Department of Science and Technology such as: (i) Facing the Facts: Women's Participation in Science, Engineering and Technology; (ii) Women in the SET workplace: Exploring the Facts, Experiences, 2005; (iii) Looking at SET through Women's Eyes; and (iv) A Monitoring and Evaluation Framework to Benchmark the Performance of Women in the National System of Innovation, 2005.

Recently, the role of information and communication technology as a tool for development has attracted a significant and sustained attention to organizations such as the United Nations. ICT access and usage can be a powerful catalyst for political and social empowerment of women, and the promotion of gender equity. While there is recognition of the potential of ICT as a tool for the promotion of gender equality and women empowerment, a 'gender gap' has also been identified and reflected in the low numbers of women accessing and using ICT compared to men.

In addition to the physical access to ICT and the ability to use it, the access also refers to the ability to make use of the information and the resources provided. Both enrolments and graduations in ICT are the lowest in the country, for both men and women as shown in Figures 19.1 and 19.2 respectively. This is alarming and it needs urgent and immediate attention. In 2005 the majority of female doctoral graduates were in the biological sciences, whereas the information, computer and communication technologies produced the least doctoral graduates for both sexes, with only 1 per cent females.

Women in ICT

One way in which gender gap and relations are challenged is through women's presence in ICT workforce. Although there have been numerous attempts by government policies and private sectors to bring more women in the ICT professions, however, women are still underrepresented and constitute a minority in the ICT professions. ICT is one of the key drivers of every nation's economy. Hence, the ICT sector has been identified as one of the key sectors by the South African government through its various national initiatives. It is evident that the access to and the use of ICT are directly linked to social and economic development of the nation. These factors together with the government efforts to promote science, engineering and technology excellence and gender equity, make imperative to understand what are the barriers that hinder women's full participation and advancement in the ICT workforce, and what can be done to remove those barriers.

The low participation of women in high-level ICT is not a problem for women only, but also for the whole world. Women miss the opportunities for technology-related job and run the risk that technological developments will not be relevant to their needs. None of countries can compete in an increasingly global ICT market if half of its talented citizens are not participating. The exclusion of women from ICTs also implies that women will have few opportunities to influence the ways in which these technologies develop and affect their lives. To increase women's participation in ICT, it is vital to establish mentoring programmes for young girls and encourage them to consider pursuing their careers in ICT. The Meraka Institute of Council for Science and Industrial Research of South African has conducted a study on participation of women in ICT. According to the South African Department of Education, the number of female graduates with ICT degree has declined between 2000 and 2004 as shown in Figure 19.3 below. A further analysis of the situation in terms of demographics is shown in Figure 19.4.

High Performance Computing (HPC)

In the past, high performance computing (HPC) was meant for academic institutions and highly specialized industries, to run data applications that carry out advanced analytics and rapid simulations. Recently, due to the advanced use of: processor power, clustering, storage technology, open source operating systems and software, the use has been extended to other sectors. HPC is a vital research application to find cure, build machines, predict weather, explore theory, discover miracle, etc. HPC delivers technologies and solutions that are accelerating scientific breakthroughs of tomorrow. In conjunction with ICT, high performance computing enables researchers to achieve remarkable breakthroughs that improve and save lives around the world. HPC is solution of speedy access to new discoveries and

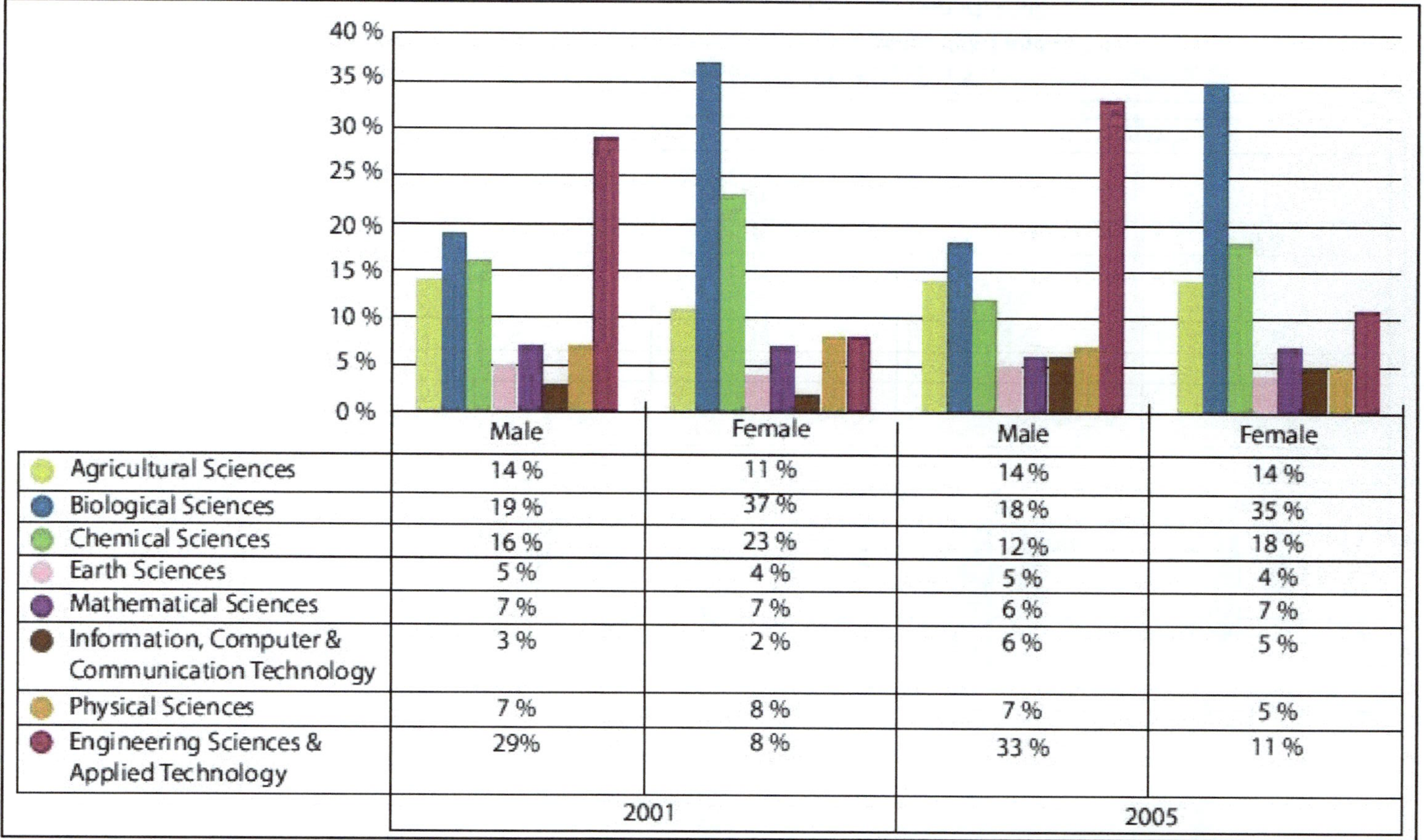

	Male	Female	Male	Female
Agricultural Sciences	14 %	11 %	14 %	14 %
Biological Sciences	19 %	37 %	18 %	35 %
Chemical Sciences	16 %	23 %	12 %	18 %
Earth Sciences	5 %	4 %	5 %	4 %
Mathematical Sciences	7 %	7 %	6 %	7 %
Information, Computer & Communication Technology	3 %	2 %	6 %	5 %
Physical Sciences	7 %	8 %	7 %	5 %
Engineering Sciences & Applied Technology	29%	8 %	33 %	11 %
	2001		2005	

Figure 19.1: South African Doctoral Enrolments in the Natural and Agricultural Sciences, Engineering and Applied Technologies Comparing Sexes (2001 and 2005).

***Source*: Department of Science and Technology Report, 2009.**

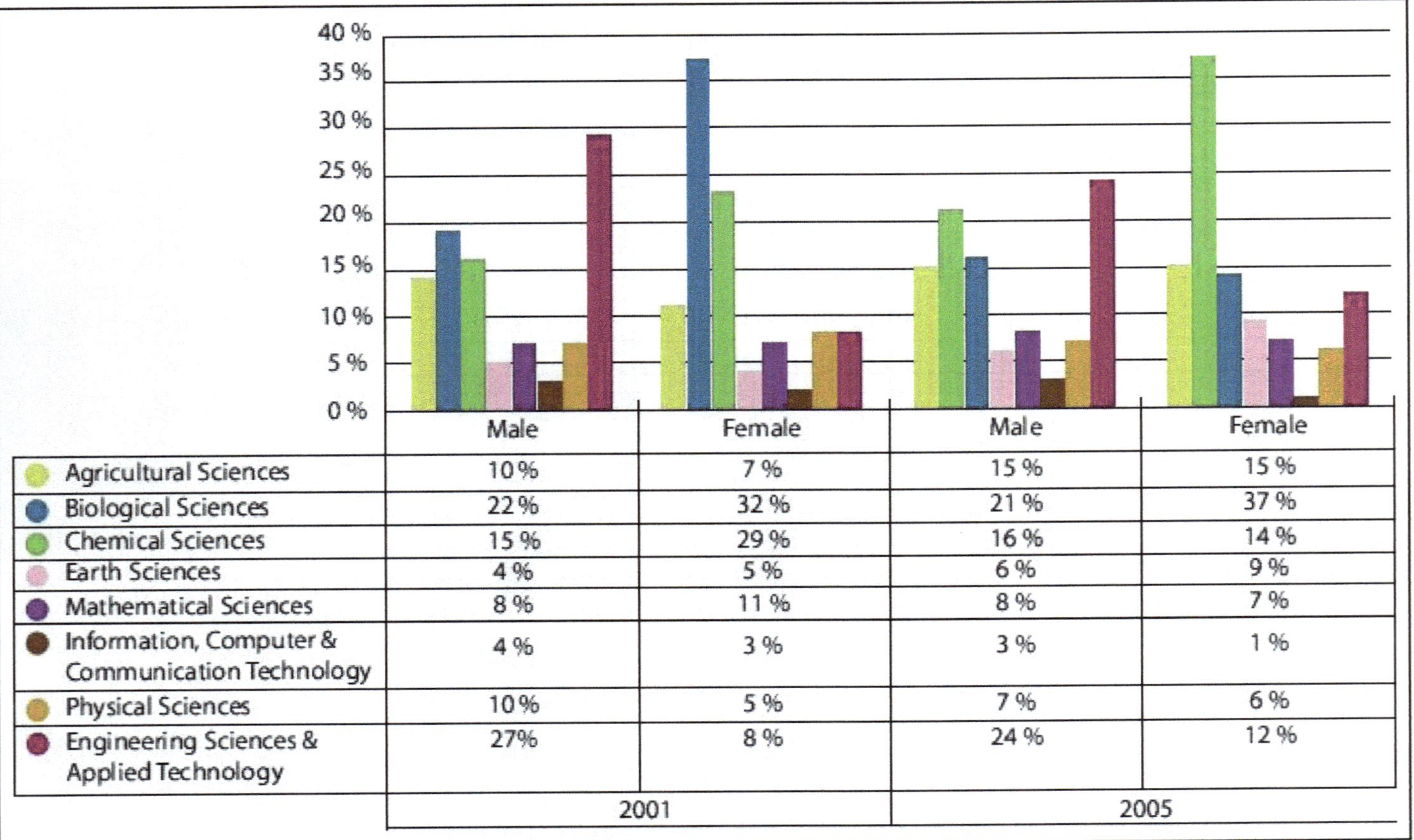

	2001 Male	2001 Female	2005 Male	2005 Female
Agricultural Sciences	10 %	7 %	15 %	15 %
Biological Sciences	22 %	32 %	21 %	37 %
Chemical Sciences	15 %	29 %	16 %	14 %
Earth Sciences	4 %	5 %	6 %	9 %
Mathematical Sciences	8 %	11 %	8 %	7 %
Information, Computer & Communication Technology	4 %	3 %	3 %	1 %
Physical Sciences	10 %	5 %	7 %	6 %
Engineering Sciences & Applied Technology	27%	8 %	24 %	12 %

Figure 19.2: South African Doctoral Graduations in the Natural and Agricultural Sciences, Engineering and Applied Technologies Comparing Sexes (2001 and 2005).

***Source*: Department of Science and Technology Report, 2009.**

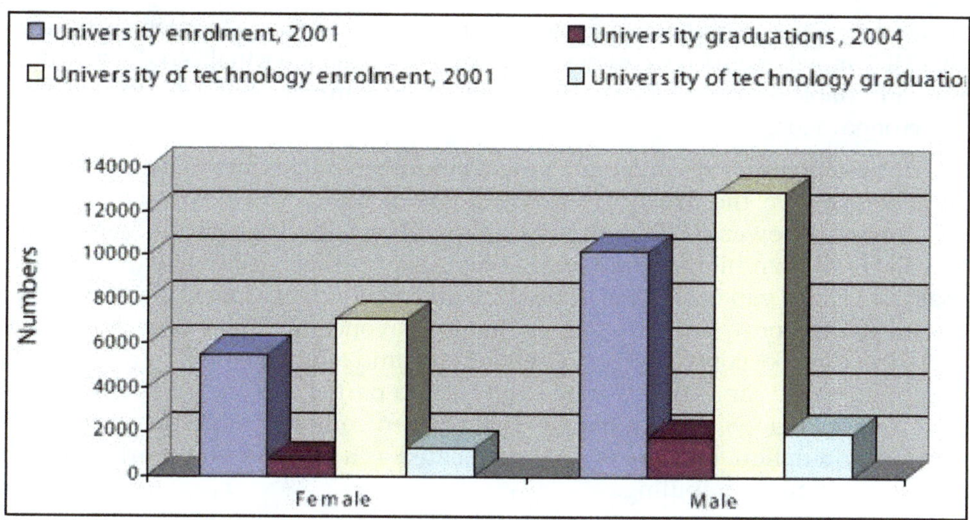

Figure 19.3: South African Graduations, Four Years after Enrolment of ICT Courses at Universities and Universities of Technology by Gender.
***Source*: The Department of Education (2000-2004).**

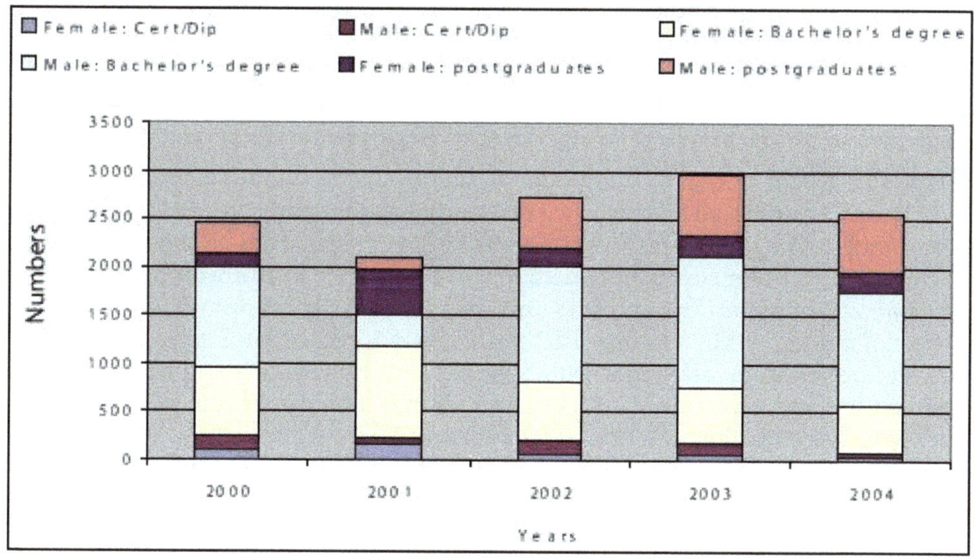

Figure 19.4: South African University ICT Qualifications by Gender.
***Source*: the Department of Education (2000-2004).**

treatments, increased research agility, lower costs and allow researchers to process, store and interpret petabytes of data.

Intensive research and development in companies are increasing, they are moving to simulations and synthesis for design and discoveries, which require HPC facilities. This is attributed in three dimensional rendering, as computational fluid dynamics, materials science, manufacturing, aeronautics, etc. Recent developments have seen that HCP moves from high cost machines to clusters and to virtualized. Thus, HPC is now affordable to organizations of various sizes. In summary, HPC is a wave of data-driven innovation.

Infrastructure is also a gender issue. There are gender issues that affect aspects of access and in the broader sense, the use of HPC. Thus, as much as there revolutionary power of HPC, the gender gap still remains. Women are, for example, less likely to own high performance computers for their research activities. This marks a gender gap in HPC infrastructure. In addition, the women's access to and use of ICT is constrained by factors that go beyond the issues of technological infrastructure. Socially constructed gender roles and relationships play a key role in determining the capacity of women and men to participate on equal terms in the information society. Despite the fact that women comprise more than half of the world's population, they are under represented as a group in ICT, particularly in high performance computing.

HPC Materials Science Research

Large scientific projects are currently conducted by using sophisticated combinations of computers, networks, visualization, data storage, remote instruments and people. Information infrastructure provides a way of integrating resources to support modern applications. In addition to the physical access of ICT and the ability to use it, the access also refers to the ability to make use of the information and the resources provided. In South Africa, there is the national HPC infrastructure, *i.e.*, the Centre for High Performance Computing (CHPC) for scientists to conduct their individual research, which is based in Cape Town.

The computational materials science has been identified as a key area of growth world widely. They become a driving force in the discovery and design of novel materials. Currently, computer simulation methods give influence to all areas of study, with a great impact in materials of physics and chemistry. With a continuously increasing power of modern HPC, sophisticated simulations and calculations are exploited for the research of innovative materials. The modern HPC is able to produce precise results for complex material systems from demanding applications. The methods such as density functional theory, *ab-initio*, molecular dynamics, energy minimization, many-body perturbation theory and quantum Monte Carlo are employed in high performance computer simulations. With the advancement of computing powers, complex materials and their properties are increasingly investigated.

The rapid and tremendous growth of materials science field has been enabled not only by developments in the theory, algorithms and software, but also by the continuing exponential growth in the power of computer hardware. There has been an improvement in the processing power from the increase of processing speed, the amount and speed of available memory of the individual processors and parallelization. There is also an increase ability to exploit massive parallelism, in which computational tasks are distributed over large numbers of processors. It is worth noting that there is a dire need for continuous improvement of the computational infrastructure (both hardware and software) for computational material science research.

The development of new materials requires an iterative sequence of synthesis and characterization; hence the high performance computing adds a new dimension to the process: materials can be synthesized and/or characterized virtually. The approach is made possible by the algorithms and hardware collectively. For example, HPC tools can be used in various advance energy technologies, such as: nuclear fission and fusion, electrochemical batteries, photovoltaic energy conversion, hydrocarbon catalysis, hydrogen storage, solar energy, etc. HPC bring together scientists (both men and women) from various fields, with one goal of using to find solutions through computing.

Conclusions

At present, a comprehensive picture of women in high performance computing across the globe does not exist. Women from various countries are attempting to map where women in HPC are, they organised workshops aimed at the location of using HPC. The observation found is that there is no sensitivity around gender issues in the high performance computing research worldwide, including South Africa. This is a global challenge that needs to be addressed. High performance computing is at the far end of ICT and almost invisible to many, hence it does not feature in gender and ICT discussions.

The overview has provided insights on the importance of high performance computing in solving the materials science research. The ICT careers, in particular, HPC research in academia is a "front-loaded" profession, which requires large investment in time and energy during the early stages of development. This is the challenge for most women as they have responsibilities of building a family and career simultaneously. Attention needs to be devoted to encourage young girls to consider pursuing careers in ICT and ultimately in HPC. Mentoring and role modelling have been shown to provide necessary encouragement tools for young girls, which in the long run will decrease the gender gap. Women should be given opportunity of comfortably use and design computer, technology, communication equipment, software and work in virtual spaces. They should not be just end users but also have to design information technologies and knowledge systems to improve life.

Acknowledgements

The author would like to acknowledge the support from the South African Department of Science and Technology.

References

1. Carvalho, T. and Santiago, R., 2008. Gender differences on research: Perceptions and use of academic time, Tertiary Education and Management, 14(4): 317-330.
2. Carvalho, T. and Santiago, R., 2010. Still academics after all. Higher Education Policy, 23(3): 397–411.
3. Denscombe, M., 1998. The good research guide. Maidenhead, Berkshire: Open University Press.

4. Department of science and technology report: www.dst.gov.za/publications/reports/sarg_booklet.pdf (accessed on 10 May 2013).
5. Department of science and technology report: Tara Research and Equity Consultants and FeedbackMetrics (24 October 2005).
6. Department of science and technology report:Centre for Research on Science and Technology (December 2005). A Monitoring and Evaluation Framework to Benchmark the Performance of Women in the NSI. University of Stellenbosch, South Africa.
7. Gardiner, M., Tiggemann, M., Kearns H. and Marshall K., 2007. Show me the money! An empirical analysis of mentoring outcomes for women in academia. Higher Education Research and Development, 26(4): 425-442.
8. Hult, C., Callister, R. R. and Sullivan, K. 2005. Is there a global warming toward women in academia? Liberal Education, 91(3): 50-57.
9. http://www.scj.go.jp/en/vision2050.pdf report for Science Council of Japan accessed (09 July 2013).
10. Kim, N., Yoon, H.J. and McLean, G.N., 2010. Policy efforts to increase women faculty in Korea: Reactions and changes at universities. Asia Pacific Education Review11(3): 285-299.
11. Lewin, T. 2006. At colleges women are leaving men in the dust, available at: http://www.uri.edu/artsci/ecn/starkey/ECN386 per cent 20-Race,Gender, per cent 20Class/girls_boys_college.pdf (Accessed on 10 July 2013).
12. Mayer, A.L. and Tikka, P.M., 2008. Family-friendly policies and gender bias in academia. *Journal of Higher Education Policy and Management*, 30(4): 363-374.
13. World Bank, 1994. Population and development: Implications for the World Bank, Washington, DC, p.75.
14. Sanderson, P. and Sommerlad, H. 2002. Gender differentiation at work: cultural capital and the conceptual articulation of structure and agency. *XV World Congress of Sociology, 7-13th July 2002*, Brisbane, Australia.
15. Theron, H. 2002. Women in academia: late starters, late achievers, available at: http://www.science-inafrica.co.za/2002/December/women.htm, site accessed 10 July 2013.
16. Townsley, N. C., and Broadfoot, K. J., 2008. Care, career, and academe: Heeding the calls of a new professoriate. Women's Studies in Communication, 31(2):132-142.

Annexure

Waknaghat Recommendations - 2013 on Empowering Women in Developing Countries through Information and Communication Technologies

WHILE EXPRESSING GRATITUDE to the JUIT Centre for Women Studies, Jaypee University of Information Technology (JUIT), the host of the International Conference on 'Empowering Women in Developing Countries through Information and Communication Technologies' in JUIT-Waknaghat, District Solan (HP), India during 1-3 June 2013;

EXPRESSING APPRECIATION to the Centre for Science and Technology of the Non-aligned and Other Developing Countries (NAM S&T Centre) for co-organising the International Conference;

EXPRESSING GRATITUDE to the Initiatives: Women In Development (IWID) and Rural Education and Development (READ) India for collaborating with the organisers of this Conference;

RECOGNISING THAT Information and Communications Technology (ICT) has an immense potential to enhance the living conditions of the people, particularly women while offering lots of opportunities and becoming a key economic tool for the developed as well as the developing countries;

WHILE EXPLORING the significance of ICT in the empowerment of women for integrating them in the technological, political and socio-economic segment;

EXTENSIVELY DELIBERATING ON the 'gender divide', effected in the lesser numbers of women accessing and using ICT compared with men; the persistent gender discrimination in labour markets, education and training opportunities, and allocation of financial resources for entrepreneurship and business development which negatively impact women's potential to fully utilise ICT for economic, social and political empowerment; and

ANALYSING the success stories of women empowerment with their involvement in various sectors while contributing to their nations' growth and development, overcoming all hurdles, putting their best to turn out to be scientists, technologists, entrepreneurs, policy and decision makers, etc;

WE, THE PARTICIPANTS OF THE INTERNATIONAL CONFERENCE, representing the scientists, academicians, professionals, managers, civil society and government policy makers of the non-aligned and other developing countries from Angola, Egypt, The Gambia, Indonesia, Iraq, Kenya, Malawi, Malaysia, Mauritius, Myanmar, Nepal, Nigeria, South Africa, Sri Lanka, Sudan, Togo, Uganda, Venezuela, Vietnam and the host country, India;

STRONGLY RECOMMEND THAT:

- The developing nations should take suitable actions against the gender divide prevalent in their society and work towards empowering women.
- The governments should take initiatives for ensuring the social security of women.
- The governments, civil society and the private sector should take initiatives for narrowing the gap between legislation and implementation of the laws concerned with security of women.
- Take measures to create awareness and educate women and civil society regarding existing laws, policies and programmes available for the protection of women.
- Mobile phones and other smart devices based applications should be developed which can help in preventing violence against women.
- Sensitisation programmes related to violence against women and its impact on society should be launched through media and ICT.
- Gender sensitive monitoring and evaluation should be regularly conducted to track the progress of women empowerment through ICT.
- Policies and mechanisms should be formulated to encourage women to enter into the areas of science, information and communication technology.
- Policy makers should be sensitised on gender issues to promote good governance.
- The girl child should be treated with dignity and nurtured with love with active efforts to change the attitude that perceives her as a burden.
- Socio-emotional empowerment of women should be promoted and training should be provided to boost their self esteem and confidence as well as develop conflict management abilities.

- ☆ Encourage and increase women's political participation and leadership at all levels.
- ☆ Allocate more resources to ensure health initiatives for girls and women.
- ☆ Policies should be formulated for ensuring technical training, its access and usage through e-education in rural areas.
- ☆ Gender stereotypes predominate and perpetuate those reflected in the media. Women's viewpoints, knowledge, experiences and concerns should be adequately reflected on the Internet and other media.
- ☆ Numerous invisible barriers limit women's participation in the information society. One of the more pervasive but intractable problems is 'techno phobia' or fear of technology. This should be taken care of by providing user friendly services.
- ☆ Women's access to and control over ICTs should be made equal to that of men. There is a need to encourage information sharing about best practices that can be replicated, while attending to local issues and social dynamics in different places.
- ☆ Policies and mechanisms should be formulated to encourage women to use ICT for income generation.
- ☆ Success stories of women should be shared through community radio, internet and other media to motivate women.

THUS, RECOMMENDED AND ADOPTED AT WAKNAGHAT, SOLAN (HP), INDIA ON THIS DAY, 3RD OF JUNE 2013.

www.ingramcontent.com/pod-product-compliance
Ingram Content Group UK Ltd.
Pitfield, Milton Keynes, MK11 3LW, UK
UKHW021010290726
14059UKWH00001BA/50